Disaster Culture

Disaster Culture

Knowledge and Uncertainty in the Wake of Human and Environmental Catastrophe

Gregory Button

Left Coast Press inc.

Walnut Creek, CA

Left Coast Press is committed to preserving ancient forests and natural resources. We elected to print this title on 30% post consumer recycled paper, processed chlorine free. As a result, for this printing, we have saved:

2 Trees (40' tall and 6-8" diameter)
1 Million BTUs of Total Energy
216 Pounds of Greenhouse Gases
1,040 Gallons of Wastewater
63 Pounds of Solid Waste

Left Coast Press made this paper choice because our printer, Thomson-Shore, Inc., is a member of Green Press Initiative, a nonprofit program dedicated to supporting authors, publishers, and suppliers in their efforts to reduce their use of fiber obtained from endangered forests.

For more information, visit www.greenpressinitiative.org

Environmental impact estimates were made using the Environmental Defense Paper Calculator. For more information visit: www.papercalculator.org.

Left Coast Press, Inc.
1630 North Main Street, #400
Walnut Creek, CA 94596
http://www.LCoastPress.com

ISBN 978-1-59874-388-3 hardcover
ISBN 978-1-59874-389-0 paperback
eISBN 978-1-59874-661-7

Library of Congress Cataloging-in-Publication Data

Button, Gregory.
  Disaster culture : knowledge and uncertainty in the wake of human and environmental catastrophe / Gregory Button.
    p. cm.
  Includes bibliographical references and index.
  ISBN 978-1-59874-388-3 (hardcover : alk. paper) — ISBN 978-1-59874-389-0 (pbk. : alk. paper)
1. Emergency management—United States. 2. Crisis management—United States. 3. Disasters—United States. 4. Uncertainty. I. Title.
  HV551.3.B88 2010
  363.340973—dc22
                                2010037435

Printed in the United States of America

⊚™ The paper used in this publication meets the minimum requirements of American National Standard for Information Sciences—Permanence of Paper for Printed Library Materials, ANSI/NISO Z39.48–1992.

Cover design by Allison Smith
Cover photograph © 1989 Natalie Fobes, all rights reserved, www.fobesphoto.com

This book is dedicated to all those who have endured
uncertainty in the wake of calamity.

Contents

Abbreviations

NOAA	National Oceanic and Atmospheric Administration
CDC	Centers for Disease Control and Prevention
CTEH	Center for Toxicology and Environmental Health
EDF	Environmental Defense Fund
EIP	Environmental Integrity Project
NIEHS	National Institute of Environmental Health Science
FEMA	Federal Emergency Management Agency
DHS	Department of Homeland Security
OSHA	Occupational Safety and Health Administration
ATSDR	Agency for Toxic Substances and Disease Registry
GAO	General Accounting Office
HUD	Housing and Urban Development
NACOSH	National Advisory Committee on Occupational Safety and Health
NIOSH	National Institute for Occupational Safety and Health
USACE	The United States Army Corps of Engineers
USGC	United States Coast Guard
USDJ	United Stated Department of Justice
USDI	United States Department of Interior
USFW	United States Fish and Wildlife
MMS	Minerals Management Services
TVA	Tennessee Valley Authority
TDEC	Tennessee Department of Environmental Conservation
TDH	Tennessee Department of Public Health
ADFG	Alaska Department of Fish and Game
ADEC	Alaska Department of Environmental Conservation
ADNR	Alaska Department of Natural Resources
NRDC	National Resource Defense Council
LDEQ	Louisiana Department of Environmental Quality
LDPH	Louisiana Department of Public Health
NYCDH	New York City Department of Health

Acknowledgments

While I have many people to thank for this book, none deserve more credit than those people who have shared their stories with me and provided me with comfort and support in the field. Unfortunately, the list is long, but you know who you are and I thank you for trusting me with your narratives and for your courage and tenacity in the face of uncertainty. In some cases the names of the people I interviewed have been changed in order to protect their privacy.

I am most grateful to friends and colleagues who over the years provided me with support and encouragement. Again, the list is long, but I would be remiss if I did not mention Robert Weiss, David Jacobson, John Petterson, Judith Zeitlin, Brinkley Messick, Benson Saler, Alan Harwood, Anthony Oliver-Smith, Barbara Rose Johnston, Monica Schoch-Spana, Carolynn Nordstrom, Alan Davis Drake, the late Paul Wellstone, and the late Roy "Skip" Rappaport.

I am also indebted to Kari Saylor Smith, Erin Eldridge, Brannon Hulsey, and especially Frankie Pack for the outstanding assistance they have provided me in preparing the manuscript.

I am most grateful to my editor Jennifer Collier for her strong support, encouragement, and expert guidance.

Most of all, I am grateful to my brother Hank, who has always been a great supporter of mine, and my wife Janet Osborn, who has loved me unfailingly and has endured my long absences in both the field and the study.

Introduction

*"The arrangements of society become most visible
when challenged by crisis."*
Eric Wolf, 1990, *American Anthropologist* 92: 586–96.

Calamity is suffused with uncertainty. In the days, weeks, months, even years that follow a disaster, people feel uncertain about real and perceived risks. Information about the nature and extent of these risks is incomplete and typically conflicting. The parties involved in the disaster, as well as organizations such as the media, public agencies, and environmental groups, release a cacophony of communications that the affected population often sees as conflicting and confusing.

Disaster victims and the general public struggle to obtain credible sources of information. In an attempt to make sense of the event, they assign meaning, blame, and responsibility and develop coping strategies. Informational uncertainty can create individual and community-wide stress and result in a lack of effective coordination between responding organizations.

Uncertainty does not simply exist—it is produced, and the production of uncertainty can result in new political, economic, and social formations. This informational uncertainty generates conflicting public discourse about blame and the responsibility for remediation. It also raises questions about the severity of the event and the potential long-term harm that may ensue. Moreover, the informational vacuum that can exist in the confusion following disasters can produce harmful rumors and misinformation. Sometimes opposing sides generate misinformation to manipulate the situation to their advantage.

In the calamity that follows a disaster, multiple uncertainties confront individuals and society. The cumulative effect can overwhelm the victims of disaster, and in the case of catastrophic events, it can overwhelm an entire society such as in the case of the recent (2010) earthquake in Haiti.

Even in a less severe disaster, citizens face stress from multiple sources of uncertainty: How can individuals continue to live and thrive after the loss of loved ones? Where will they live now that their home has been destroyed? How will they be able to survive without income? How have their lives been irretrievably altered by the loss of their social support networks? If they are forcibly displaced, how will they live as strangers in a strange land? How will they and their loved ones cope with psychological trauma; every time they have a cold or the flu, how will they know if they are presenting the onset of illness resulting from toxic exposure? How can they overcome the uncertainty over their environment? In short, how can their lives return to normal?

The Role of Science

Since science and technology are often involved in disasters, people often instinctively turn to scientists for answers, as do politicians, corporations, and public agencies. For example, following most unnatural disasters, conclusive scientific evidence about both the cause and nature of the catastrophe and the potential deleterious effects of exposure to toxic hazards is usually lacking. While the media seeks immediate answers to meet daily deadlines and the public demands instant answers, science, by virtue of its methodologically rigorous approach, cannot readily respond to these demands. Science can be a slow and painstaking process that cannot often produce the sound bites that the media demands or the reassurance or insights the public craves. Moreover, some uncertainties cannot be adequately addressed because sufficient scientific information does not exist (Nelkin, 1985: 18).

For instance, often no clearly calibrated safe exposure levels to toxic hazards exist. If such exposure levels do exist, their accuracy is often debatable. Usually, specific correlations with diseases, for example certain types of cancer, are unknown. As Devra Davis has astutely observed, "epidemiology is a blunt instrument" (2002). While there is, for example, sufficient epidemiological evidence to prove that radiation in some cases causes various forms of cancer, there is no scientific way to establish whether radiation exposure has caused a specific case of leukemia. Inconsistent and contradictory scientific studies of the etiology of disease are often the result of imprecise understanding of disease. Under these circumstances, corporations can all too easily deny a disease-exposure relationship to toxic illness.

Usually, such an informational vacuum generates a climate of controversy both within the scientific community as well as with the residents who may have been exposed to toxic substances. Lay people are sometimes confused by the ambiguous or contradictory statements made by experts and puzzled about how to make sense of the ambiguities involved.

On the other hand, experts are often perplexed by the concerns expressed by disaster victims and their preoccupation with fear and uncertainty. Consequently, the two sides begin talking past one another. In such an uncertain atmosphere exacerbated by conflicting worldviews, citizens often become skeptical about the reliability of scientific evidence. Suddenly, some see the scientific discipline as lacking in certainty and incapable of resolving ambiguity. In the process, science's systematic invincibility is called into question, and its monopoly on truth is challenged. A dispute between lay and expert erupts. Whose voices are heard and whose voices are denied become essential to the contestation over the meaning attached to the disaster. Among those whose lives are disrupted by tragedy, the production of unhidden transcripts emerges during this period of conflict.

The Role of Corporations

Lay people are not the only ones who become disillusioned with science in the context of disaster. Fearing that litigation and increased regulatory control may threaten their profit-making activities, corporations demand a sounder science in order to influence public opinion, the media, and government agencies. The goal of transforming science and restraining regulation does not, ironically, result in the lessening of uncertainty but sometimes in the increased production of uncertainty. In this climate of scientific uncertainty, corporations cast doubt on their critics and seek to undermine challenges to the *official* narrative and attempt to avoid the scrutiny of government agencies. Corporations often resort to defining the event in a more limited scope in order to prevent government interference. Often they dismiss claims of critics by labeling them as junk science, "irrational," or "overly emotional."

Policymakers are then faced with the task of creating and implementing policy in a highly charged arena in which evidence-based policy is severely constrained and hotly contested. The climate of uncertainty is sometimes mutually constituted when government agencies seek to maintain the asymmetrical balance of power by protecting corporate and state interests over the public good. In some cases, government agencies seek to avoid the scrutiny of the public and other government agencies and either amplify uncertainty or withhold vital information.

Beneath the contestation over the status of science lurks the larger question: Just how much should we allow science to influence the political process? If science is to play a role, we need to ask "whose science" (Harding, 1991; Wing, 1998).

In this vortex of uncertainty and absence of knowledge, various interest groups turn to the expertise within the community. Amid all the conflicting claims and accounts, irreconcilable evidence and pervasive ambiguities, conflicting truths and differing canonical views of the event become polarized, which can either clarify the situation or accentuate the uncertainty. More often than not, the latter occurs.

The position the layperson and expert assume in these disputes is influenced not only by empirical facts but also by special interests, personal values, and the cultural and moral implications of a given technology. Very often opponents are embedded in their opposition to, or support of, various technologies and belief systems over and above scientific evidence.

The Production of Uncertainty

What becomes apparent in these discursive practices is that uncertainty is just as endemic in science as in other realms of life. This recognition has increasingly been employed as a tactic to create further uncertainty. Thus, in recent decades, uncertainty has become politically inflected. As we shall see, many corporations and public agencies involved in these uncertainties adopt a public relations strategy and policies that attempt to create either a sense of certainty in order to reassure and downplay the disaster, or, as a political strategy, contribute to a climate of amplified uncertainty.

On occasion, advocacy groups have also employed similar strategies, though not to date to the degree employed by corporations and the state. In such instances, science is often shaped, skewed, distorted, manipulated, or manufactured in order to retard liability or protect polluters from government regulation or cast blame and responsibility on others. Such tactics not only manipulate science and create a smoke-screen of uncertainty but also serve to undermine our public regulatory systems. Consequently, citizens are made vulnerable to future hazards and our system of governance is undermined.

The Failure to Recognize the Important Role of Uncertainty

In recent decades, a robust and expansive literature on the nature and role of risk in modern day society has emerged (Knight, 1921; Beck, 1992; 1999; Giddens, 1998). This preoccupation has largely relegated the dynamic category of "uncertainty" to a black box. Too often uncertainty is trivialized. Risk discourse can be viewed, at least in part, as an attempt to overcome or lessen uncertainty and thereby protect citizens from future hazards of all kinds.

Although uncertainty is an essential part of the equation in mitigating risk, it has been neglected because admittedly it does not, at least from a rationalist perspective, lend itself easily to analysis. By definition, uncertainty is elusive and does not lend itself to the same scientific calculus that examines the nature of risk. Moreover, one cannot easily contemplate relative uncertainty in the way that relative risk can be analyzed. The notion evades the scrutiny of stochastic models and the technical calculations employed by most risk analysts.

Thus, uncertainty is too often relegated to the realm of the irrational. At times it is dismissed by analysts, policy makers, and politicians as some kind of free-floating anxiety whose pursuit is fruitless and threatens to undermine rational discourse. Ironically, some are reluctant to admit that this discourse is shaped by political, economic, bureaucratic, ideological, and cultural concerns.

The Need to Examine Uncertainty in a New Light

Uncertainty, because of its elusiveness, seems irrational. Its appearance as an internal rather than external threat seems most susceptible to psycho-cognitive analysis. Nonetheless, to limit our exploration to this realm alone and ignore the sociocultural, political, and economic forces that also foster uncertainty would be a mistake. Despite uncertainty's expansive and elusive nature, common denominators about the nature of uncertainty can be readily observed as well as how this seemingly elusive category is being increasingly manufactured.

Introduction

Uncertainty is susceptible to exploration and categorization if viewed as lived-experience and, more important, as being socially and culturally constructed. The nature of uncertainty can be explored by employing an ethnographic rather than a rationalistic approach. This approach can unveil the social, cultural, and political meanings inherent in the nature of uncertainty, which are too often ignored. By contextualizing uncertainty in the domains of culture, meaning, and power, we create a coherent idea of its role in society. Thereby, we call into question what is normative and what imaginative alternatives might be constructed.

This book, based on three decades of disaster research, is primarily concerned with the production of uncertainty as an ideological tactic. My goal is to examine how the conscious tendency to manufacture, revise, or withhold knowledge politicizes the discourse in the wake of disasters. In previous publications (1995; 1999; 2002), I have focused on how the media packages information and participates in the construction of reality, thereby informing us of some of the ideological elements that seek to maintain the status quo in the wake of crisis. Such packaging also reminds us of the often-overlooked fact that disasters are not merely phenomena in the material world. They are grounded in the politically powerful world of social relations. The analytical approach I employ underscores that disasters are not only socially and physically disruptive, they are also political events. Thus, disaster analysis must examine power relations among the various agencies and institutions involved in the event and the people affected by disaster (Button, 2001).

Whether labeled "natural" or "unnatural," disasters highlight the asymmetrical distribution of power and foreground the struggle of the state, corporations, and human agency for the redistribution of power. The control of information in public discourse, as well as the attempt to control the social production of meaning, is an attempt to define reality. It is, therefore, a distinctly ideological process that we cannot afford to ignore. I view uncertainty as a prism for studying the social formations of disaster and seek to address the question of how politically generated uncertainty reconfigures both the landscape of disaster and our social arrangements. In the process, it inflicts greater harm in the wake of disasters and makes effective response difficult.

A Different Way in Which to Examine Disasters

This book provides a comparative analysis of the ways in which the politics of uncertainty is inflected in a time of calamity. By examining selected case studies based on my three decades of fieldwork research on disasters ranging from Love Canal, Three Mile Island, toxic groundwater contamination in Woburn, Massachusetts, the destruction of the 9/11 Twin Towers to the *Exxon-Valdez* oil spill, Hurricane Katrina, and the TVA ash spill, I attempt to unpack the significance of the nature of uncertainty in a time of calamity and propose a new analytical grid to understand the dynamic nature of disasters.

My approach is to employ ethnographic case studies to inform my analysis. However, I use the term "case" somewhat cautiously since disasters are not isolated events bound by space, time, or even concept, but rather are set in motion by a set of preconditioned series events and are followed by a series of cascading events that continue to unfold over time (Oliver-Smith, 2009). We usually perceive disasters as isolated and abnormal events. Disasters, like all social processes, are moving targets, which we can temporally isolate but not fully capture, in our textual analyses and more often than not our attempts to isolate them result in oversimplification. I employ a long-term comparative analysis in which we can distinguish disasters as routine and normal, connected to one another along various social fault lines and a direct product of our culture, not something to be imagined as simply an exceptional event. If we conceive of them in this sense they do not represent a traditional bounded unit of analysis that typifies normal case studies. Like Fortun's (2001) observation on Bhopal, these and all manifestations of categories we label "disasters" are not bounded, but belong to a larger set of spiraling events. Thus, what I refer to as "case studies" in this instance are more like freeze frames of crisis, momentarily captured in time and space simply for the ease of analysis. Like all disasters, they continue to unfold across all traditional boundaries and across time. In some sense, disasters are similar to the way in which Clifford and Marcus (1986: 18) describe cultures, "contested, temporal, and emergent." Hopefully, these case studies and the comparative analysis that follows can become useful stepping-stones for rendering a more holistic account of the political dynamics of uncertainty in times of calamity.

Each of the six ethnographic accounts of disasters has something specific to say about the nature of uncertainty in times of calamity. At the same time, the six examples share common recurrent themes that tell us on a general scale about how we as a culture respond to disasters. The six chapters are followed by four chapters that provide an analytical approach by which to investigate the production of uncertainty and the sequestration of knowledge.

Following this introduction, I provide accounts of the *Exxon-Valdez* oil spill, which introduces some of the primary themes of the book, as well as informs the reader about the nature of that specific event. Each of the case studies that follow demonstrates the lessons learned, or not learned, from the different disasters while demonstrating recurrent patterns that show how uncertainty and limited access to knowledge play formative roles in our culture's response to disaster.

The seventh case study on the BP Deepwater spill was added as this book was being prepared for publication. It not only serves to underscore dramatically the recurrent themes of the book, but also presents the persistent harm inflicted by the manufacture of uncertainty up to the present day. I hope that this new tragic chapter in our history, which is positioned in a long line of similar tragedies, will force us to reexamine underlying causes of this spiraling series of events.

In the analytical chapters that follow the case histories, I examine some of the perduring themes of the book, such as mediated accounts of disasters, the contestation over knowledge, the production of uncertainty, and the sequestration of knowledge in journalistic, scientific, legal, governmental, and corporate domains. These chapters also introduce broader accounts of disasters with the intent of demonstrating how endemic the production of uncertainty is in our society.

1. A Sea of Uncertainty

When I first heard of the spill, I didn't think it would affect me or anyone in the Homer area. I felt sorry for the fishermen in Prince William Sound and I felt sorry for the animals and the environment since I had hunted and fished in the sound many times over the years. However, as the spill came around the outer coast and neared Kachemak Bay I began to realize that it was going to profoundly affect my life and the lives of all animals and humans in the bay. My crew and I went out on my ship to see the oiled beaches. In order to get some idea of what to expect. Nothing prepared us for what we saw. There was total destruction. It was as if an atomic bomb had gone off. Crude oil was all over the beach up to the high tide area. There were countless dead birds littered about the beach and we heard the screams of a dying baby otter as it lay next to its dead mother. I couldn't take it. I sank down to my knees and sobbed uncontrollably. We were all crying. It was then that I realized our paradise was in jeopardy. We decided to return to Homer and tell the others what we had seen. Everything about our future felt terribly uncertain.

Homer Fisherman (Stout, July 14, 1990)

On March 24, 1989, the *Exxon-Valdez* oil tanker, a ship longer than three football fields and carrying 1,286,738 barrels of crude oil, went aground on Bligh Reef in Prince William Sound, Alaska. In the largest oil spill in

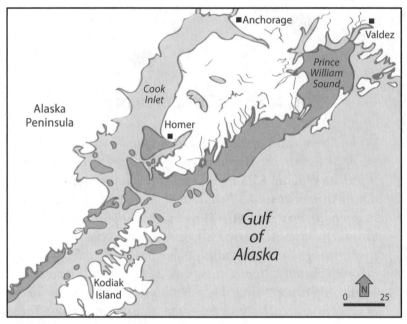

Map of the Alaska Oil Spill.

North American history, eleven million gallons of crude oil flowed into a pristine wilderness area and affected twenty-four coastal communities. Within a week, the spill would cover 900 square miles of water; within two weeks, 1,300 miles were contaminated. Homer, Alaska, is a small city on the shore of Kachemak Bay on the Kenai Peninsula. Residents of the town, which is 530 road-miles from the city of Valdez, were saddened and outraged but never suspected that the spill would soil their own waters in Cook Inlet and Kachemak Bay. Like other Alaskans, they were worried about the wildlife in Prince William Sound and were concerned about the fate of the fishing industry. Many Homer residents held permits to fish in the Sound, and the Homer cannery processed over 1,000 tons of fish from the Sound annually.

It soon became the benchmark in U.S. environmental history by which all ensuing environmental disasters were to be measured. The spill acted as a catalyst for the environmental movement just as the Santa Barbara oil spill (1969) had two decades earlier. The impact of the spill far exceeded any harm previously inflicted by an oil spill in U.S. waters. The impact on the environment was massive. More than 350,000 sea birds were killed, along with 144 bald eagles, 5,500 sea otters, thirty seals, and twenty-two whales. These mortality statistics are based only on the animal corpses recovered. Some scientists suspect that nearly half the corpses of marine mammals were lost at sea and never recovered. Sharp declines in pink salmon and herring in Prince William Sound have been associated with the devastation of the marine habitat. The economic and sociocultural impacts were also severe. Both short- and long-term harm was inflicted on individuals, families, businesses, local governments, and Native Alaskan communities. Spill-related social conflict and spill-related disruption of social routines were frequently reported. Threats to traditional subsistence and cultures were most common in native communities. One major study reported that "because of how the clean-up was organized and because of how the money was spent, the socioeconomic and psychological damage to the communities worsened" (Impact Assessment, 1990; Button, 1993).

Soon, the first of many uncertainties began to unfold. In the days following the spill, the National Oceanographic and Atmospheric Administration (NOAA) assured Alaskans that little of the oil would leave Prince William Sound. A few fishermen who hailed from New England recalled the false assurances made by government scientists in the wake of the wreck of the *Amoco Cadiz* in 1976 that the oil would never come ashore. They were among the first to be wary of NOAA's and Exxon's claims (Winslow, 1978). As Homer residents watched the tides, there was growing skepticism about assurances that the oil would not harm their waters. Their intimate and extensive knowledge of the tides in Alaskan waters told them otherwise. In mounting anger and distrust, they began to challenge NOAA and Exxon on their forecast. They urged precautionary measures that would protect coastal areas outside of Prince William Sound. The skepticism of residents was well founded, based not only on their keen understanding of the local tides but also on earlier scientific research that suggested that the state of the art of oil spill movement "was not without its uncertainties." An article in *Technology Review* stated, "We still do not understand how the waves passing underneath an oil slick, the wind blowing over it, and the gross motions of the underlying water combine to move oil on the surface of the seas" (1976).

In spite of the locals' knowledge of the complex tides in Prince William Sound and the Gulf of Alaska, NOAA scientists and Exxon officials dismissed the concerns of coastal residents, as they would for years to come. Later, an NOAA official would tell me that he had "learned a valuable lesson" and that in the future, in addition to computer modeling, he would give more credence to local knowledge (NOAA Official, July 10, 1990).

Six days after the spill, a notice appeared in the *Homer News* warning that the oil could reach Kachemak Bay. The next day, the oil escaped from Prince William Sound and threatened the outer coast of the Kenai Peninsula and the fishing waters of Homer area residents. Slowly, the oil crept toward the Homer coastline.

As the oil spread, concerns grew about the potential impact on local ways of life. Perhaps most poignant were the Alaskan Native responses. After some villagers in Port Graham returned from the outer coast with news of the oil's arrival there, the women of Port Graham went out that night to the outer coast at low tide to gather chitons (small mollusks) from the rocks. After cooking and cleaning them, they distributed one plastic bag of these much-prized delicacies to every household in the village (Kearnes, 1990). As the natives of English Bay observed the inevitable advance of crude oil, they, too, began gathering food from the intertidal zone. Like the villagers in Port Graham, English Bay residents especially sought out the mollusks. Robert Kvasnikoff, the village corporation president, said, "It's like the Good Lord told you, you have one more hour to live. Go see your friends one last time" (April 23, 1990).

Putting the Fox in Charge of the Chicken Coop

In an unanticipated move, President George Bush Sr. placed Exxon in charge of the oil spill cleanup. Many Alaskans, both pro- and anti-oil, were stunned and furious about this decision. Many said it was tantamount to putting a fox to guard the chicken coop or putting a rapist in charge of helping his victim. For some, it violated ideals of American democracy. One Kodiak resident said, "In a democracy, the party responsible for negligence or criminal offences isn't given power to decide how restitution is to be made, what it is, when it is completed, and where it is accomplished" (TDTWD, 1990).

According to the oil spill response contingency plans, Alyeska, the consortium of seven oil companies that operates the pipeline from the North Slope of Alaska to Prince William Sound, was supposed to be legally responsible for oil spill cleanups. British Petroleum (BP) had the majority holding and so bore most of the responsibility for the spill response, even though it was an Exxon tanker that ran aground. As the primary partner in the consortium, BP made most of the decisions in the initial days following the spill, decisions that allowed the spill to swell out of control and become the largest oil spill in U.S. waters in history.

Alyeska's oil spill contingency plan was found wanting. Among other errors, it overlooked, according to sociologist Lee Clarke, the qualitative difference between small and very large spills (1999: 107). Alyeska's plans for a spill of great magnitude were grossly inadequate to the challenge of responding to the Exxon spill. Its contingency report stated: "Alyeska believes it is highly unlikely a spill of this magnitude (200,000 barrels) would occur" (Alyeska, 1987: 3.5). Their scenario for a large spill included "seas less than five feet, currents less than 1.6 knots, waves less than 2 feet, 2 miles visibility, and winds of 5 knots" (Alyeska, 1987). These almost perfect conditions were highly optimistic for Prince William Sound. In keeping with this optimism, the report's authors stated that booms and skimmers would arrive at the spill within two and one-half to five hours (U.S. Senate, April 6, 1989). The response plan made the common error—or used the common strategy—of relying on a best-case, rather than a worst-case, scenario. Furthermore, as with all contingency plans, it made promises it could not keep.

The response effort was crippled from the beginning. The spill response station that was supposed to be erected in the Sound was never built. Instead of having large quantities of boom on hand, only enough was available to surround the tanker. The timeliness of the response was also severely curtailed by the fact that the response barge, which was supposed to be in harbor and fully stocked for immediate deployment at all times, was, instead, on shore, buried under snow, and ill-equipped to respond. The skimmers that were supposed to be part of the response effort were out of commission. It took seven hours for Alyeska's helicopter to arrive at the spill site. Not only was the oil spill response equipment not in a state of readiness, the oil spill response team had been disbanded.

The plan originally called for cleaning up oil spills in Prince William Sound efficiently enough as to preclude the need for a state or federal takeover of the effort. It called for a five-hour response time and estimated that 50 percent of the oil would be recovered. This was an overly optimistic projection: typically, only 10 to 15 percent of spilled oil is recovered (U. S. GAO, October, 1989). In Prince William Sound, the recovery was less than 10 percent—probably somewhere between 6 and 8 percent (U. S. Congress, Office of Technology Assessment, 1991).

The State of Alaska also had an oil spill contingency plan. However, the Alaska Department of Environmental Conservation did not have a full-time response team or maintain a large inventory of response equipment and supplies—nor did it have a plan for a spill anywhere near the magnitude of the *Exxon-Valdez* spill. The plan was voluntary; it was vague and contained no language regarding how the state would interact with either Alyeska or federal response teams.

Had all of these response systems been in place, the *Exxon-Valdez* might have remained a small chapter in the history of oil spills. However, the contingency plans were largely blue smoke and mirrors. Valdez resident Mike Lewis summed up the feelings of many Alaskans:

> The failure in the Prince William Sound oil spill was a failure of technology and it was a triumph of public relations. We had been promised this wonderful shining example of technology, with the marine terminal facility and pipeline, and the ability to handle these tankers through Prince William Sound. And then on March 24th we suddenly woke up with the realization that what had been happening all along was that technology was never put in place, but millions of dollars were spent on public relations to make everybody think they were okay. (Citizens Commission Hearings, November 1989: 14)

As it turned out, Lewis's interpretation of how the spill response unfolded is not far from the truth. An anonymous source within Exxon told me that Exxon brought in their public relations team first and their logistics team last. This decision apparently infuriated members of the logistics team, who were ready to go within ten hours after the tanker went aground. Later, in an interview with *Fortune* magazine, a statement by Exxon CEO Lawrence Rawl revealed this focus on public relations: "You'd better pre think which way you are going to jump from a public relations standpoint before you have any kind of problem. You ought to always have a public relations plan, even though it is kind of hard to force yourself to think of a chemical plant blowing up or spilling all that oil in Prince William Sound" (Davidson, 1999: 206). Paradoxically, despite Rawl's statement and all the money Exxon spent to improve its public image, the company made many PR mistakes, especially in the early days of the spill. For example, for several days Rawls refused to be interviewed by the press or go to Alaska. Matters worsened when, on April 3, 1989, Exxon released "An Open Letter to the Public," which was printed in over one hundred major media outlets. The letter, signed by Rawls, stated, "We believe we acted swiftly and competently to minimize the effects this oil will have on the environment, fish, and other wildlife." This stiffly worded apology backfired and infuriated many people because, by the time this letter was issued, it was already an accepted *fact* that Exxon had acted slowly and incompetently. Many people in Alaska to whom I spoke were also incensed that the letter neglected to mention the impact the oil would have on their lives as well as the environment. Insensitive gaffes like this proved to be harmful to BP's CEO, as we shall see in a later chapter.

When Governor Steve Cowper learned of the delays and inadequacies of the response plans, he requested that President Bush federalize the spill. Alaska senators Ted Stevens and Frank Murkowski supported this request. Admiral Nelson of the Coast Guard refused to federalize the spill, saying the arrangement would be too cumbersome. U.S. Coast Guard Admiral Yost explained in a Senate hearing that the federal government was afraid that, if such as step were taken, Exxon might withhold money for the cleanup and the federal government would have to pay for it (U. S. Senate, April 6, 1989: 75–76). If the Coast Guard merely coordinated the cleanup, however, Exxon would continue "to keep their checkbook open." Under Yost's scenario, "we basically direct and guide Exxon in where we want the cleanup, where we want the skimmers." Unfortunately, Yost assumed the Coast Guard would have more authority to direct than it ended up having, an assumption that would haunt him in the coming months. The Tanker Owners Pollution Federation was also critical of the plan. The Federation argued that although the polluter was responsible for the cleanup, it lacked authority to respond. Other organizations were also critical of placing Exxon in charge.

The Petroleum Industry Response Organization argued that large spills such the *Exxon-Valdez* spill should be federalized in order to insure an unambiguous command structure (U. S. Congress, Office of Technology Assessment, 1990). The Natural Resources Defense Council (1991) also joined the fray and argued that, not only should large oil spills be federalized immediately in order to insure a clear command structure, but such a plan would have the public interest as its foremost consideration.

Even the Office of Technology Assessment was leery of the plan to place Exxon in charge. In light of the *Exxon-Valdez* oil spill, they argued that, "by the time the Coast Guard determines the polluter is incapable of dealing with a spill, it may well be too late for anyone to mount a successful countermeasure effort" (U. S. Congress, Office of Technology Assessment, 1990).

Ultimately, President Bush ignored these pleas and decided against federalizing the spill and asked the Coast Guard to coordinate and monitor Exxon's effort.

The National Response Team Report to the President (May 1989) was critical of Alyeska, Exxon, the state, and the federal government for not being prepared for a spill the size of the *Exxon Valdez* spill. There was considerable confusion between Alyeska and Exxon in the immediate hours after the spill because neither of their contingency plans had clear guidelines for coordinating a spill response. When it was determined that Alyeska was not prepared to respond, Exxon assumed control of the cleanup; however, both organizations failed to inform the state or the federal government of this decision. This move created greater confusion and seriously impeded communications between Exxon and the state and federal response teams.

As the National Response Team pointed out (May, 1989: 12), the shortage of equipment "delayed any opportunity to contain the spilled oil early on." The failure of Alyeska to respond immediately to the spill and the revelations that the state, federal, and Exxon oil spill contingency plans were inadequate undermined the public's belief in the credibility of all parties. Alyeska's failures instilled a strong distrust of authorities, which permeated the cleanup activities for the duration of the clean up over the next two years.

Some Homer residents reported they were furious at what they saw as *collusion* between the federal government and Exxon, while state and local officials apparently stood idly by. Some residents were upset because they thought that the Coast Guard and the oil industry had a too cozy relationship. Some informants alleged that there was a revolving door between the two in which retired Coast Guard officials were hired by the oil industry. Later, the Alaska Oil Spill Commission Report (1989) was quite critical of the Coast Guard and stated that it had "an unduly friendly relationship with [the oil] industry." The report accused the Coast Guard of being lax in its enforcement of tanker safety: "The Coast Guard has failed the American People in providing oversight of the country's oil transportation system."

Larry Smith, a twenty-year resident of Homer and an environmental activist, wished for a governor like George Wallace, who would have, metaphorically, blocked the school house door and stood up for state's rights: "Governor Cowper should have gone over to Valdez and shut down the pipeline until the oil industry responded in a responsible way to the spill" (Smith, August 1, 1990). Homer's mayor, John Calhoun, accurately summed up local views in his testimony at the Alaska Oil Spill Commission Hearings in the middle of July:

> The Coast Guard certainly could tell Exxon what to do. But if Exxon wanted to flip them off, they could do that, what kind of citation could the Coast Guard write them? I mean they've [Exxon] got more people sitting back there in law offices than they've got cleaning up the oil spill, they'll tell the Admiral right where to send it and smile. (AOSCH, 7/15/89: 29)

Mei Mei Evans, a Homer resident, at the same hearing, expressed the rage and frustration of many citizens:

> While Exxon continues to play games of rhetoric and posturing with federal, state and local representatives, the oil continues to contaminate our shoreline and shut down our fisheries. As recently as yesterday afternoon, Exxon's latest public relations representative demonstrated that corporation's continuing insensitivity when he cheerfully observed to the Homer multi-agency advisory committee that "only two percent" of the total Alaska coastline has been affected

by the spill. ... There is a fundamental moral contradiction inherent in this situation, and that contradiction continues to abrade local residents every day. The spiller was given control of the restoration, and to this day, one hundred and fourteen days after the tanker went aground, Exxon continues to call the shots in the clean up process. When have we ever before allowed the perpetrator of the crime to determine the course of the rehabilitation? Do you put an arsonist in charge of extinguishing the fire?—or a rapist in charge of rendering aid to the victim? (AOSCH, July 15, 1989)

Uncertainties Surrounding the Remedial Cleanup Methods

In the wake of the spill, there was a cascading series of uncertainties: uncertainty about what communities and waters would be affected; uncertainty (even among scientists) about the appropriate cleanup methods. Uncertainty and lack of consensus about appropriate, safe, and effective methods divided not only people within the affected communities, but also people from various state and federal agencies, the U.S. Coast Guard, and scientists within Exxon.

Both regulatory agencies and the oil industry approached the oil spill cleanup problems as a set of scientific and technological problems that were to be solved by a rational decision-making process. In contrast, many of the Homer-area residents did not see them in this light. For instance, those who opposed Exxon's spraying of the beaches with fertilizers grounded their opposition on the belief that huge multinationals should not be making decisions about local environmental concerns. Since the remedial technologies proposed by Exxon assumed varying degrees of risk, some local people wanted to choose the risks rather than have them imposed by either the government or corporate entities.

Many locals also argued that both the agencies and Exxon had vested interests in selecting various cleanup techniques. It was, for instance, commonly rumored that Exxon preferred using Inipol (a cleanup chemical) because it owned the French company that manufactured the chemical.

While Exxon's strategies seemed, at least on the surface, rational, Homer-area residents could not view these decisions in merely an objective, detached, and rational manner. Both their livelihoods and the environment (to which they were deeply attached) hung in the balance, and they distrusted bureaucratic decisions being made in Washington, D.C. or Houston, Texas (Exxon's headquarters). In essence, many locals felt they bore a disproportionate level of risk in the face of the threat from the oil spill and did not want outsiders making decisions that affected their lives. One local fisherman told me that "since Exxon and the federal agencies in bed with them caused this problem we shouldn't trust them to fix it" (Stockpole, June 7, 1990).

The highly politicized environment made the handling of uncertainty all the more problematic. Since both science and technology played a central role in the response to the oil spill, they were fertile ground for conflicting perceptions about risks. The nature of the debates about appropriate technological responses was so complex and entertained so many uncertainties that there was considerable disagreement among scientists as well as among lay people, politicians, residents, and Exxon employees.

Dispersants

One of the most controversial issues was the use of chemical dispersants in cleaning up the oil. Dispersants are solvents used to break down the cohesiveness of the oil that are sprayed on a spill in order to remove the oil from the surface of the water. The chemical agents cause the oil to enter the water column in tiny droplets. The oil droplets are then dispersed in the water column and become diluted, ostensibly until they are in such low concentrations that they are considered harmless to the environment. Early dispersants were considered toxic; at the time of the *Exxon-Valdez* spill, some were considered less toxic than other dispersants (National Research Council, 1989). A good deal of the controversy over the use of dispersants, as well as the controversy over the efficacy of dispersants, had to do with the concern for their impact on the environment.

Using dispersants has primary biological advantages. They can reduce the hazard of oil spills to birds, that is, unless the birds come in contact with the dispersants. Not only are dispersants often toxic, but their chemicals can destroy the water repellency and insulating capacity of bird feathers (National Resources Defense Council, 1990). Some argued that an advantage to using dispersants is that breaking up the oil prevents it from fouling shorelines, which are extremely difficult to clean (U. S. Congress, Office of Technology Assessment, 1990). At least, that was the thought at the time. Today, some scientists believe that dispersants can have potentially harmful effects on coastlines.

Ever since the sinking of the tanker *Torrey Canyon* off the coast of Great Britain in 1967, dispersants have had a bad reputation. Many observers at the time thought that the dispersants used by the British government in the wake of the *Torrey Canyon* going aground caused more harm than good to the marine ecosystem and the bird population (Cowan, 1968: 108; Easton, 1972: 21). Even today, twenty-one years after the *Exxon-Valdez* oil spill, the use of dispersants is still a highly controversial topic.

Homer residents cited numerous studies that the toxic effects of the chemicals did more harm than good (National Response Team, May 1989). They argued that large populations of algae, limpets, barnacles, and mussels were destroyed by the chemical compounds. The effectiveness of dispersants was largely unknown at the time and is still hotly disputed. Even though by the time of the *Exxon-Valdez* spill they had been by then used on more than fifty spills, studies of their effectiveness were undercut by lack of controls and poor documentation.

The Natural Resource Defense Council (1990) contended that the ineffectiveness of dispersants had only been demonstrated in laboratory tests and planned field tests. The council argued that "contrary to popular perception," dispersants do not remove oil from the ocean: "They merely transfer it from one locale—the ocean surface—to another, the water column. The impact is simply shifted and not eliminated" (NRDC, 1990: 31). This is an argument that many still make today. Nevertheless, the National Research Council approved the use of dispersants and recommended that they be considered as a first response measure (National Response Team Report, 1989: 21) while at the same time stressing the need to conduct more research on both the effectiveness of dispersants and the potential harm.

The Alaskan Regional Response Team had previously, in its response plan, authorized the use of dispersants in Alaskan waters in order to reduce response time but according to strict guidelines. The team established guidelines for Prince William Sound by dividing it into three different zones. In zone 1, dispersants were preauthorized for use by the on-scene coordinator. In zone 2 the use was conditional on protecting natural resources. The on-scene coordinator was required to submit a formal request for their use, subject to the approval of both the Environmental Protection Agency (EPA) and the Alaska Department of Environmental Conservation. In zone 3, where the oil had drifted by the second day, the use of dispersants was not recommended, and any use required a formal proposal similar to the above procedure.

Exxon requested permission to apply dispersants the second day. However, the company had neither sufficient quantities on hand nor the planes to disperse them. The oil spill response contingency plan, which Exxon shared with the North Slope oil spill consortium, of which BP was the major player, failed to have both the chemicals and planes on site in the Port of Valdez as they had promised in their plans. Instead, they only had 350 barrels in three sites in Alaska, which was only enough to clean about 9 percent of the spill. In fact, there were not enough dispersants anywhere in the world for a spill the size of the *Exxon-Valdez*. Even though Exxon's own scientists argued against the use of dispersants, Exxon ordered its plants in Houston and England to manufacture more immediately (Davidson, 1990).

Thirty-six hours after the spill, Exxon attempted to apply an experimental trial of dispersants to the spill in zones 1 and 2, adjacent to the tanker. Other than spraying the tanker crew and Coast Guard members, the operation had very little effect. Despite this failure, Exxon then applied for approval to spray dispersants on Knight Island Passage (zone 3). The state denied the request, noting the failures of the previous attempts. This denial and high seas and storm conditions prevented Exxon from trying again. L. G. Rawl, chairman of the board and CEO of Exxon Corporation, testified in a U.S. Senate hearing that the delay in approval cost the corporation its best opportunity to mitigate the spill (U. S. Senate, April 6, 1989). Some observers speculated that Rawl's statement was a strategic move partially motivated by legal considerations based on a French judge's decision in the *Amoco Cadiz* spill. It had been argued that the French government had not allowed the use of dispersants in that instance (Frost, March 27, 1990).

Burning Oil

In situ burning is another controversial oil spill cleanup technique. As with dispersants, it requires an environmental trade-off. A burn-off of spilled oil potentially could get rid of 90 percent of the oil (U. S. Congress, Office of Technology Assessment, 1990) and prevent shoreline oiling. However, scientists disagree about the environmental costs of burning. Some contend that harmful toxins are released in the air by this method. Toxic compounds, including aromatic hydrocarbons, evaporate after a spill. While lighter compounds, such as benzene and toluene, are released into the air quickly, heavier compounds, naphthalene for instance, remain on the surface. Fewer of these light hydrocarbons are released when the oil is burned, but the heavier and more toxic ones are released into the air (Scientific American, 1991). One report on in situ burning disagrees with these findings. According to the Office of Technology Assessment, in a report written by David Evans for the National Institute of Standards and Technology, the compounds released by the burning are no more harmful than those that would normally evaporate. However, a report by the Natural Resource Defense Council (1990) disputes this report and claims that burning releases toxic compounds. Furthermore, it points out that when burning oil slicks it is impossible to contain the vapors or control where they go. The report does admit that there may be times when the trade-off of burning is worth the risk. The Office of Technology and Assessment (1990) states that the burning of oil creates a highly visible column of smoke that raises several concerns about human health effects, whereas leaving the oil on the surface of the water is generally perceived by the public to be less of a threat to human health. Despite the health concerns, the Alaska Department of Environmental Conservation did not oppose other burnings as long as nearby residents were given adequate warning. However, the weather by the third day made burning an impossible option. Exxon went ahead with the controlled burning without providing adequate warnings, a decision that would further undermine the company's probity in the eyes of the public.

"Collateral Damage"

The native village of Tatitlek was alarmed when a sooty, black cloud of smoke entered their village the second day of the spill. Exxon was burning somewhere between 12,000 and 15,000 gallons of oil (The National Response Team Report, 1989), but the villagers, five miles downwind, were not notified of the experimental burn. Village residents reported eye, nose, and throat irritation and were concerned for pregnant women and the elderly. The native residents with whom I spoke reported that they were "incensed," "angered," and "mad" that no one had notified them and that no one had taken the health hazard into account. One resident told me that Exxon's action demonstrated to him and his neighbors that the oil company had a callous disregard for Alaskan Natives.

Toxic Contamination of the Food Chain

The concerns of the residents of Tatitlek about the burning of oil were not the only concerns of Native Alaskans. They, and other Alaskan residents, were worried about the threat to their subsistence way of life, a way of life that is heavily dependent on the marine ecosystem. From the initial days of the spill, natives in every village reported either suspicious conditions in the subsistence foodstuff or unusual animal behavior (Fall, 1990). These conditions aroused concerns about the advisability of eating what they perceived as contaminated food. They argued for tests on the tissue of potentially contaminated species in their food chain. Nonnative subsistence users raised similar concerns. Convinced they were being given the run-around, coastal Alaskans demanded a thorough response from health officials. A health task force was created to address these concerns.

Once established, the health task force admitted, "a virtual absence of studies on human health effects, direct, or indirect, existed on the possible transfer of contaminants through the food chain to man" (Health Task Force, March, 1990). One toxicologist told me that, although a number of occupational health and safety studies on the potential toxicity of crude oil existed, there were virtually no studies on the potential toxicity of oil in the food chain.

The task force conducted several tests. Unfortunately, the results of the first test were not released until nearly five months after the spill, near the end of the harvest season. The health task force published a newsletter that made it quite clear, in response to questions posed by natives, how little was actually known and just how much uncertainty existed. The following example well illustrates this:

Q: Are there levels of aromatic hydrocarbons that are considered safe for human consumption?

A: There are no established acceptable levels of aromatic hydrocarbons in food. Aromatic hydrocarbons are ubiquitous. They are present in many foods consumed including cooked and smoked meats, fish, grains and cereal products, and fruits and vegetables.

Q: Is there a way to test if people have eaten oil-contaminated food?

A: There are no feasible tests that are available to test for or monitor human exposure to aromatic hydrocarbons or other components of crude oil.

Q: Are there specific ways that concerned Alaskans can reduce their potential risks?

A: Any risk of adverse health effects as a result of the oil spill from the consumption of Alaskan fish and shellfish is very small. Villagers should rely on common sense and their own judgment to avoid collecting foods from areas obviously impacted from the oil. In addition, individuals should decide based on appearance, smell, texture and taste of subsistence foods. If food is of doubtful quality it should not be consumed. (Health Task Force, March, 1990)

This latter statement greatly frustrated many people. In a public meeting with health officials, one native elder made an astute point in this regard. As he rose to speak, he pointed out that although he had only a grade school education, this statement seemed illogical to him. A few moments earlier, a health spokesperson had stated that the toxicity of crude oil existed in levels of parts-per-billion. If this is the case, the elderly man stated, how could the safety of food based on appearance or smell be judged since parts per billion would unlikely affect either? The health task force members were immediately taken aback by the sense of the speaker's question. Another villager asked the quite logical question: "If experts really don't know about toxicity levels of aromatic compounds, how can you make the statement that the risk is small?"

Frequently, natives complained that environmental tests should be conducted in more locations and in a broader range of places (Fall, 1990). Since Exxon was a member of the Health Task Force, many people treated the task force statements with skepticism. Because scientific experiments and collections were made jointly with NOAA and Exxon, many informants were highly skeptical of the process in which Exxon participated. As we shall see in a later chapter, these concerns resemble concerns about the Tennessee Valley Authority's joint research with state and federal agencies in the wake of the TVA ash spill disaster in 2009 as well as other joint ventures between the government and polluters.

Worker Health and Safety

Other health concerns slowly began to emerge among cleanup workers and their families. Some of these concerns persist until the present. Very early in the cleanup, workers began to complain of headaches, sore throats, persistent coughs, runny noses, respiratory problems, and flu-like symptoms. Some workers became disabled and had to stay at home. Other workers didn't manifest more serious symptoms until years later. According to the National Institute for Occupational Safety and Health (NIOSH), the number one non-injury worker compensation claim made during the first year of the cleanup (1989) was for respiratory damage (Reller). Some years later, a discovery effort in a court case brought to light Exxon's internal medical reports that stated that in the first summer "an unspecified number of the 11,000 workers made 5,600 clinic visits" for an unspecified respiratory illness (Phillips).

Many of the 15,000 cleanup workers were working twelve-hour days, fourteen days at a stretch in cold, wet conditions. The high-pressure steam hoses used to wash the oil off rocks created a crude oil mist that was aerosolized, further exposing them to risk. The risks to the workers were not unknown. Crude oil contains a number of toxic chemicals such as benzene, toluene, and polynuclear aromatic hydrocarbons, or what is commonly referred to as PAH (a group of one hundred compounds, some of which cause cancer). While toluene and benzene evaporate quickly, PAHs and solvents posed some of the greatest risks.

Exxon and its contractors had told the workers that they were in no danger working around the crude oil and using the chemicals that were employed in the cleanup. They were told that the only precautions they had to take were to avoid skin contact and other common-sense precautions. The State of Alaska also took the position that there was minimal risk for the workers. An advisory issued by the state's Department of Health and Social Services stated there was no adverse health effects from breathing the air in contaminated areas. While it recognized there might be some risks to cleanup workers, the risks were minimal and could be prevented by worker training and the proper protective equipment. This statement came only after the state's Department of Labor declared the cleanup a hazardous waste operation. However, the declaration of a hazardous waste operation did not come without a struggle.

Union inspectors, with funding from the National Institute of Environmental Sciences, expressed concerns about inadequate worker health and safety training and the dangers of inhaling or ingesting crude oil and being exposed to other chemicals, including solvents. Workers complained that there were not enough respirators available. Some workers who did have access to respirators stopped using them. They found them difficult to wear in the humid conditions and also difficult to wear for as long as a twelve-hour shift. Other workers claimed that respirators, gloves, and other protective gear were not made available until several weeks after the cleanup began. In a deposition, one cleanup worker reported in a congressional oversight hearing that when workers asked why VECO, Exxon's subcontractor responsible for cleanup, could not provide them with respirators, a VECO foreman told the workers to go home if they didn't want to work without respirators.

According to a report in the *Los Angeles Times*, a Coast Guard safety officer stated that "nobody complied with any of the health and safety rules, and everybody turned a blind eye. They were issuing rain suits [as protective gear] and a rain suit is [worthless] as protective equipment except for one chemical: water" (Murphy, November 5, 2001).

Eula Bingham, who was the assistant labor secretary for Occupational Safety and Health Administration (OSHA) in the Carter Administration, and who was the leader of the laborers team, complained that investigating worker safety proved to be very difficult. In her opinion, Exxon did not cooperate. There was also the added obstacle (that others including the press encountered) that getting to remote sites required assistance from Exxon. Exxon and its subcontractor had leased almost all the available planes, helicopters, and boats in the state so people who required assistance in getting around were often beholding to Exxon. Bingham said, "I must say, we got the runaround. I am sure [Exxon] didn't want me around. It has always troubled me over the years" (Ott, 2005).

Troubled by what they saw, the team listed risks to workers to include skin cancer (from dermal contact), leukemia (from inhalation of benzene), respiratory damage (from the use of high-pressured steam hoses to clean contaminated rocks), and potentially long-term or chronic effects that could include blood and liver disorders and damage to the central nervous system. Bingham pressured the state to declare the cleanup a hazardous waste operation, which it eventually did, thereby requiring health monitoring on site and training in hazardous waste cleaning.

OSHA standards in such conditions normally require workers to have forty hours of training, but the state ignored this standard and allowed Exxon, at its request, to conduct only four hours of training. Many critics later alleged that the training did not provide adequate protection, and some even suggested that the training violated federal right-to-know provisions by not fully informing the workers of the toxicity of the chemicals involved. An NIOSH report, however, stated that the training was adequate and "no major omissions in subject matter or problems with delivery were noted." Curiously, though, the report also stated that, "there is always room for improvement in course content or the manner in which the information is presented" (NIOSH, May, 1991).

Two other statements in the report are worth noting. In the conclusion, the report notes that the wearing of protective gear "was not consistently enforced from one work site to the next and that "the hands and forearms of many workers were contaminated with 'weathered' crude oil" (NIOSH, May, 1991). Another telling comment is that "there is a need to develop and coordinate an injury/illness surveillance system as soon as possible after the work begins" (NIOSH, May, 1991). Ironically, the authors noted that, while most of the *Exxon-Valdez* spill cleanup was concluded (that is in the first year), this and other recommendations were being made about the conduct of future oil spill cleanups. In other words, they were admitting indirectly that there was no surveillance in the first year of the cleanup when the majority of workers were exposed to potentially the greatest harm. Furthermore, in retrospect, there is no adequate way to assess the impact on worker injury and illness.

Carl Reller, who was then VECO's environmental manager, was upset with the federal officials caving into Exxon's demands: "The decision was based on a conservative premise that was not revisited. Was this because of legitimate oversight, incompetence, conspiracy, cost cutting or negligence? Based on my experience I would say all of the above" (Murphy, November 5, 2001).

NIOSH tried to conduct its own study of worker exposure and had very limited success. Mitch Singal, a NIOSH officer, expressed frustration over trying to obtain data from Exxon: "The company refused to open the records. ... We had a right to the data under the Occupational Safety and Health Act of 1970." The NIOSH investigator's report simply states that there was "an unsuccessful attempt to conduct a systematic, record based review of health and injury data in the field" (May, 1991). NIOSH was forced to gather its own data, which proved to be inconclusive. Some of the air samples detected benzene but at levels, according to NIOSH, not great enough to cause harm. Other tests suggested that some workers might have been exposed to elevated levels of diesel fumes (which can cause cancer).

Another study, after the cleanup was over, was more successful in gaining access to Exxon data. Reller later joined the Alaska Health Project and conducted a follow-up study at the request of attorneys pursuing an action against EXXON on behalf of a former worker. The project used over 6,000 air samples collected for Exxon by Med-Tox Associates. Examining these records, he was able to evaluate the occupational exposure to oil mist during the remediation period of the spill. In evaluating the health and safety implications of employing high pressure hot water and steam to clean up the beaches, he concluded that monitoring records reveal that the average oil mist exposure for workers exceeded the NIOSH's permissible exposure limit (PEL) twelve times. He also found "serious problems" with

Exxon's laboratory procedures and data interpretation. During the spill, the data collected by Med-Tox Associates was not available to the public. However, as a result of law suits in the following years, some of the data became accessible.

According to court records, Med-Tox only conducted thirty tests for the long-lasting, more dangerous, threats such as PAH, but 1,600 for light hydrocarbons, such as benzene, which are known to evaporate within seventy-two hours and probably posed little threat to most workers. Reller and others have also criticized the monitoring of toxins because toxic standards are usually based on the average eight-hour workday and forty-hour workweek. Cleanup workers often worked twelve hours a day and for two or more weeks straight. Thus, according to Reller, the Med-Tox studies failed to adjust the allowable exposures to reflect the prolonged exposure rate. Equally troubling is Reller's claim that Med-Tox's reliance on the standard reference material for evaluating risk for exposure to oil mist was inadequate and misleading. Med-Tox relied on NIOSH's standard reference material: mineral oil. Reller argues that this standard is inadequate because mineral oil is a "highly purified product designed for non-toxicity" whereas the cleanup workers were exposed to Prudhoe Bay crude oil (PBCO), which is highly toxic and contains numerous highly toxic compounds. Given the difference between mineral oil and PBCO, Reller argues the laboratory equipment used to evaluate PEL should have been calibrated using PBCO—not mineral oil—as a standard.

The Coast Guard in its final report (1993), in a statement that proved to be prophetic, noted that many unanswered questions about worker safety may not be fully addressed for years: "The matter is likely to remain unresolved for some time, and worker health issues may ultimately be litigated, perhaps in significant numbers."

Bioremediation of the Spill Site

Bioremediation is another nonmechanical cleanup technology that met with much controversy and also brought into question the possibility of deleterious long-term health effects. Oil degrades over time after it has been broken down by both sunlight and microbes. Bioremediation is an attempt to accelerate this process. The process is simply the application, in situ, of microbes to biodegrade and oxidize hydrocarbons. Bioremediation is not a new process; it has been used for a long time as a routine landfill treatment and at hazardous waste sites. It has seldom been used for oil spills. However, at these kinds of sites, scientists can regulate temperature, nutrients, and oxygen (Scientific American, 1991).

In August 1989, Exxon and the EPA began experimenting with this technique on approximately seventy miles of shoreline. These beach areas were covered with two kinds of nitrogen and phosphorus-bearing fertilizers (U. S. Congress, Office of Technology Assessment, 1990). Fertilizer is thought to enhance the growth of oil-eating bacteria. So far the tests are inconclusive, and they have not been without controversy. Exxon has maintained that the application of fertilizer has accelerated the rates of biodegradation somewhere between fivefold and tenfold, whereas, the EPA has stated the enhanced rates range from twofold to fivefold. Other researchers have stated even the EPA's conservative figures are wishful thinking in that there is no statistical difference between the beaches treated with fertilizer and those that were not treated (Scientific American, 1991). Biochemist Don Button has argued that there was no need for applying fertilizer because the seawater in the sound contains plenty of nitrogen and phosphorus. Ronald Atlas, a microbiologist, has stated that because of the abundance of microbes, Prince William Sound is not an ideal test site for bioremediation: "If you already have a high rate of degradation, it is hard to show enhanced degradation" (Scientific American, 1991).

Concerns about Using Fertilizer Inipol

Other more serious questions have arisen. Exxon used a fertilizer product called Inipol EAP 22. It is oleophilic: that is, it sticks to oil and contains oleic acid, which is a source of carbon. According to studies conducted by the Canadian Department of Fisheries, Inipol was found to slow the process of hydrocarbon degradation. The researchers concluded that the microorganisms preferred to eat the Inipol rather than the oil. Questions have also been raised about the harm to marine organisms. Anecdotal reports suggest that beaches sprayed with Inipol had increased growths of filamentous green and filamentous brown algae. This effect has not been reported by EPA or Exxon scientists. Exxon also claims that in only one of four tests for toxicity to marine animals did Inipol prove to be harmful. The chemical adversely affected oyster larvae eighteen hours after the application. Moreover, the chemical poses serious health problems for humans. Workers learned that breathing the fumes could cause internal bleeding. Some workers had blood in their urine (according to the Seldovian health clinic physician); only later were workers informed of the dangers (Alaska Public Radio, September, 16, 1989).

Sandy Alsas, VECO supervisor in Seldovia, stated that about eighteen Seldovian cleanup workers were exposed to Inipol, but they were never told about its health effects. Workers experienced burns, headaches, and nausea. She said after several days of pleading by the workers, VECO finally sent someone to conduct urine and blood tests. Alsas said that after the laboratory tests, VECO admitted that the workers may have been exposed to bioremediation chemicals. The crew was immediately laid off, and according to Alsas, VECO never commented again on the incident (March 31, 1990).

Recently, more serious questions have been raised about Inipol. Claiming serious illnesses resulting from an exposure to both crude oil and Inipol, several people have filed claims against Exxon and VECO. Claimants state they are suffering from shortness of breath, dizziness, skin lesions, headaches, and neurological disorders. Dr. Rea, a Dallas physician who specializes in treating victims of petroleum-related poisonings, attributes the illness of his patients to a synergistic combination of toxic fumes from crude oil and Inipol. The doctor alleges that one patient has already died and that some of his other patients are going to die as well.

The *Boston Globe*, which reported the cases, obtained a copy of an Exxon company document that states that Inipol is carcinogenic and, if inhaled, can cause "dizziness, headaches, respiratory irritation, and possibly death." The document is also reported to have contained a warning from OSHA that exposure to Inipol may cause "eye and skin irritation and blood and kidney damage" (Coughlin, May 10, 1992). Exxon's material data safety sheet states, among other things, that "health studies have shown that petroleum hydrocarbons and synthetic lubricants pose potential health risks, which may vary from person to person. As a precaution, exposure to liquids, vapors, mists or fumes should be minimized. Inhalation of high vapor concentrations may have results ranging from dizziness, headache, and respiratory irritation to unconsciousness and death" (Coughlin, May 10, 1992).

Claims by these former workers and Dr. Rea are bound to be surrounded by uncertainty. Exposure to the alleged chemical(s) may or may not be easily proved. Even if exposure can be documented, it may be difficult to prove that the reported illnesses are directly attributable to the chemical(s) in question. Furthermore, unless there has been considerable medical research about the chemical(s) there will no doubt be uncertainty about the diagnosis, prognosis, and treatment. Moreover, former workers who think they may have been exposed to the chemical(s) in question may have to live with the uncertainty of whether or not such exposure will compromise the quality of their life or will even kill them until evidence that is more concrete is forthcoming.

Over the last two decades since the spill, some former cleanup workers have continued to complain of health problems. High profile legal advocates like Erin Brockovich and attorney Ed Masey have been involved in the controversy. Riki Ott, a marine toxicologist from Cordova, Alaska, has been instrumental since the early days of the spill in keeping the issue of worker health and safety alive (Ott, 2005). Her perseverance has insured that the worker health and safety remain a central concern in spills since 1989.

Fewer than two dozen former workers have filed suits against Exxon and VECO. Eight of the suits have been dismissed by the courts, and seven were settled out of court. Exxon and VECO spokespeople have denied that their cleanup efforts were responsible for any systemic illness patterns and insist that they adhered to worker safety standards. Riki Ott has written a compelling and extensive book examining the allegations. She documents the accounts of a number of workers who have had lingering symptoms since their participation in the clean up (Ott, 2005).

Limited Access to Information

Several weeks after the spill, at Exxon's request, the U.S. Ninth Circuit Court ordered a rule of silence on any issue connected in any way with oil spill litigation. The order addressed all research activity that was supported by state and federal agencies or by Exxon projects. Although the order did not affect independent research, of which there was little to speak of, it did affect approximately sixty research projects, thus making it impossible for anyone to obtain scientific information. The order raised a series of objections as well as disturbing questions. These questions included concerns about the privatization of science and concerns about government science becoming politicized. It raised a red flag about the validity of scientific research becoming connected with litigation, especially in regard to the fact that because the science was sequestered it was not available for the scrutiny of peer review. Scientists also raised questions about the order constraining academic freedom as well as the need for research data to be available to address similar problems without having to wait several years before it was made public (Davidson, 1999: 203–4). As we shall see in later chapters, these questions became the focus of controversy and debate in future disasters including the TVA ash spill (2008) and the BP Deepwater spill (2010).

How Do You Define "Clean"?

Exxon judged the cleanup a success. It was anything but that. An ongoing controversy questions, "When is a beach clean?" At first, Exxon officials promised to clean all beaches, but their definition of "clean" changed over the course of the coming months. The fishermen and the state insisted on a definition of zero tolerance. They wanted all the oil to be removed without a trace remaining. Exxon changed its objective semantics from "clean" to "treated." When asked what they meant, one Exxon person defined it this way: "Gross contamination had been removed and the shoreline left in an environmentally stable position, meaning it is not harmful to the natural inhabitants of that shoreline" (Davidson, 1982: 216). State and federal officials, reporters and environmentalists wanted to know exactly what Exxon meant by this statement. Otto Harrison, the company's chief representative in Alaska, explained that Exxon (unilaterally) had redefined its goals. "We are not removing all the oil, but removing enough so the environment can stabilize and restore itself" (Davidson, 1982: 185).

An example of this policy was Knight Island in Prince William Sound. Exxon touted the Island as the worst place left unclean. However, it argued that it wanted to leave the beach oiled because, in the company's opinion, removing the oil "would do more harm than good" (Frost, 1990: A4). The Alaska Department of Environmental Conservation did not understand and was not happy with this redefinition. Ray Bane, superintendent of Katmai National Park, expressed concern about the new definition: "Now they don't say clean, they say 'environmentally stable.' That's bull. A corpse is environmentally stable" (Dumanoski, 1989). On the other hand, the state's policy of zero tolerance was impractical, and Harrison's charge that such a policy was a marketing decision for the state's fisheries and not an environmental decision had some merit. Nevertheless, the controversy stemmed from the fact that some people interpreted Exxon's redefinition to be a backing off from their original commitment and a semantic justification for leaving beaches with oil on them.

Exxon's cleanup for the Homer area ended September 15, 1989. After the demobilization of Exxon/VECO cleanup crews, controversy erupted over whether the U.S. Coast Guard should have signed off on beaches that allegedly were not clean. According to a report in the *Homer News*, the Coast Guard signed off on more than a third of the regional beaches over the objections of both local and state authorities. *Unanimous* opposition to the Coast Guard signing off on twenty-seven beaches was reported. Further controversy arose because allegedly Exxon was reported to have *routinely* claimed to have cleaned beaches on which no work had been conducted. According to the newspaper account, an investigation of the records revealed that Exxon claimed it cleaned seventeen beaches on which, according to its own records, the company had never worked (Ortega, September 21, 1989).

Exxon seemed more confident about its cleanup methods and the power of *mother nature* to restore the environment. Lawrence Rawls, Exxon CEO, stated publicly that in ten years there would be no evidence of the spill or remaining damage. In contrast, of 596 respondents in a household survey conducted in eleven communities affected by the spill, 74.1 percent were less optimistic. Almost three-quarters of the randomly selected respondents stated that they thought the effects were permanent or would last at least as long as the rest of their lives (Impact Assessment, 1990: xiv). Because no clearly-defined, mutually-agreed upon goal at the outset of the cleanup was set, the efforts ended in a mess of ambiguity and controversy.

Two decades later, crude oil is still present in Alaskan waters and on beaches. According to a statement by the *Exxon Valdez* Oil Spill Council, the oil is decreasing at a rate of 0–4 percent per year. By these calculations, it will take many more decades, possibly a century, to disappear. Over the years, I have witnessed many beaches and inlets that have remained oiled, as far away from Prince William Sound as Kodiak Island. Every time I return to Alaska, someone presents me with a bottle of oil gathered at a beach or insists on taking me to an oiled inlet to demonstrate the oil's persistence in the environment. One middle-aged Alaskan friend is worried that the oil will not disappear in his lifetime and may still be present when his grandchildren are adults.

Epilogue

The *Exxon-Valdez* oil spill not only called into question our nation's oil policies but re-galvanized the environmental movement, even more so than the Santa Barbara oil spill two decades earlier. Perhaps for the first time since the Great Dust Bowl of the 1930s, it reminded the nation the degree to which an environmental disaster could inflict long-term harm on human communities as well as the natural environment.

The spill also placed in stark relief the many uncertainties that are inherent in catastrophes. Questions emerged about what constitutes safe and effective remedies while, at the same time, demonstrated once again that, once a major amount of oil is released into the environment, it is very difficult, if not impossible, to control it, let alone remove it from our seas and shores. If nothing else, it demonstrated once again that our technological sophistication for responding to spills lags far behind our technological innovation in locating and capturing petroleum. Furthermore, the spill raised disturbing questions, which have been raised in the past, about our nation and culture's sociopolitical ability not only to prevent disasters but also to respond to them effectively.

We witnessed, perhaps more clearly than in the past, how polluters attempt to downplay a disaster and rely more on public relations strategies than earnestly trying to undo the harm that their negligence has caused. Certainly, the grossly inadequate oil spill response plans of Alyeska, Exxon, and the state and federal governments demonstrated that cost-cutting efforts and unrealistic notions about worst-case scenarios make us more vulnerable to the ravages of disaster. The spill also raised troubling questions about the cozy relationship between government and the private sector, questions that also serve to illustrate the lack of transparency within both the government and corporate sectors.

Additionally, the spill highlighted questions and uncertainties about the contamination of the food chain and equally disturbing questions about worker health and safety. These concerns were obscured by Exxon's withholding of information and by science's and industry's continuing failure to investigate the short- and long-term health effects and the intricacies of worker health and safety in the wake of oil spills.

Beyond these issues, as we shall see in the chapters that follow, the *Exxon-Valdez* spill brings to the fore a number of questions about uncertainty. How do corporations, state agencies, social advocacy organizations, and other actors attempt to control disaster narratives? How do they adopt public relations strategies that either downplay or amplify a sense of uncertainty in order to advance political goals?

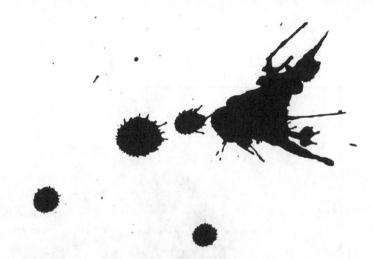

2. Uncertainty and Social Conflict over Animal Rescue

A mother otter and her pup lie dead in a black pool of oil, their large, childlike eyes full of innocence and suffering. Most people remember this kind of image most when they think of the *Exxon-Valdez* oil spill. The pup seems to represent the embodiment of innocence; the dead mother, a tragic reminder of our failure to protect and guard the creatures of our *last great wilderness*. Such images poured into the living rooms of millions of people around the globe. Newspapers, periodicals, scientific journals, news magazines, and televisions replayed these striking images again and again. In the first six months following the spill, major evening news showed eighty images of oiled otters. The photographs sold newspapers, magazines, and airtime.

For me, the most memorable image of the spill is not an anthropomorphized otter, but a bucket of crude oil with bird feet of different sizes and shapes haphazardly jutting out from the surface. At first glance, the feet appear as twigs stuck in a gooey mess. They are anonymous deaths that symbolize the devastation of the spill better than the romantic images more commonly displayed. They remind me of the skeletal leg bones of corpses piled up on a battlefield. Indeed, scenes of animal and environmental destruction can be powerfully evocative. In the aftermath of the Santa Barbara oil spill, a *Life* photographer who had viewed a hundred dead sea lions and elephant seals said that, although he had photographed disasters and wars around the world, the sight of these dead animals was the most sickening he had witnessed (Easton, 1972: 168). Another photojournalist covering the Exxon spill stated she thought it was "her obligation

Five weeks after the Exxon Valdez ran aground in Prince William Sound, fishermen net an oiled otter near Green Island. The otter was taken to the animal rescue facility in Valdez, Alaska.
Photograph © Natalie Fobes, www.fobesphoto.com.

to get these images in the living rooms of the nation and to keep them there as long as possible" (Keeble, 1991: 169). Thousands of people from all over the United States and the world were touched by these pathetic sights. Some came to Alaska at their own expense to work as volunteers at the animal centers. An exploration of the responses to the deaths of these animals gives us an ideal opportunity to probe the interstices between ideas, symbols, and culture and contestation over the meaning of an event.

Symbolic Importance of Animals

The death of animals holds several layers of meaning for our discussion of the disaster. Animals took on an important symbolic value for both Alaskans and non-Alaskans, but the deaths also had tremendous psychological and social consequences. Many people I spoke with experienced depression, loss of sleep, and other symptoms of bereavement, though no human lives had been lost. Some of them recognized this as a grieving process. The depth and extent of these experiences indicate a need for consideration of the serious results of nonhuman loss of life in the wake of a major disaster.

Various informants and groups experienced and reacted to these losses in different ways. Natives and nonnatives related differently to these animals before the disaster, and these prior relations conditioned their responses to the suffering and deaths of these nonhuman creatures. In addition, divisions developed within the nonnative community, most notably between the scientists and the animal rights activists involved in rescue and rehabilitation efforts. Each of these groups expressed very different value systems, each linked to their worldviews. These values often came directly into conflict, leading to social controversies of serious proportions. Intermingled with these differing worldviews were controversy and conflict about the proper role of science and uncertainty surrounding what was the best way to deal with the plight of the otters. The clash between lay and scientific worldviews and the uncertainty of differing approaches provide us with a unique opportunity to explore scientific uncertainty in the wake of disasters.

The attitudes and controversies surrounding this conflict are not necessarily directly attributable to the spill or cleanup issues per se, but rather to preexisting values about animal life and the role of science. Each group proved adept at managing symbolic images for their own purposes. Sometimes, this manipulation grew directly from their value systems; at other times, it may have stemmed from material self-interest. Determining the line between value-driven and politically driven interests is difficult, if not impossible.

Massive animal casualties have great symbolic importance for all of us, both Alaskans and those distant from the tragedy, as they have in disasters that both preceded the *Exxon-Valdez* oil spill and those that followed. The massive deaths of animals following the *Torrey Canyon* oil spill (1967), the Santa Barbara spill (1969), polybrominated biphenyl (PBB) contamination of cattle in Michigan (1973), the *Amoco Cadiz* oil spill (1978), the Shetland Islands oil spill (1993), the *Sea Empress* oil spill (1995), the nuclear meltdown in Chernobyl (1989), and Hurricane Katrina (2005) all resulted in public concern about the impact of these events on animals.

For some, the deaths of animals from the *Exxon-Valdez* spill represented the death of innocents, the death of a once pristine environment. Such a tragedy possibly served as a harbinger of our own destruction due to unchecked, environmental degradation. However, for some people these dead also represented resurrection and even prophecy. One Homer-area informant, a biologist, told me that the spill was, in effect, the "second coming of Christ." According to this intelligent and highly articulate woman, the suffering of the animals and their death represented Christ's suffering and death for the sins of men. Some religious informants made an analogy between the death of the animals and the death of the sacrificial lamb in Christian theology. Catholic, Protestant, and Russian Orthodox sectarians attributed significance to the fact that the spill occurred on Good Friday. Roger Wharton, a minister of the Church of Holy Trinity in Juneau, preached: "God's wonderful creatures gave their lives so that we might learn from their deaths that life on this planet is threatened to the core. ... Three days after this spill it was Easter Sunday but there was no resurrection for the animals killed as the death continued to spread through Prince William Sound and the food chain" (O'Meara, 1989: 71). In Anchorage a month after the spill, a religious service attended by hundreds of people was held at Westchester Lagoon in order to observe five minutes of silence for the animals. On some beaches, tiny crosses were stuck in the rocks or etched in the oil to mark the site of fallen animals.

Both the corporations responsible for the spill and the environmentalists critical of Exxon seemed to employ these images to their own ends: both sought to manage these powerfully evocative images. As Roderick Nash has trenchantly observed, environmentalists in the twentieth century have come to employ photographs in a manner reminiscent of nineteenth century abolitionists (1989: 208). Images of slaves branded, tortured, whipped, and boiled in brine were used by abolitionists to alter public opinion. So, too, have sea otters become a contested image. Some Greenpeace activists have realized the potency of these images. According to a Homer man who was involved in the incident, a group of activists attempted to dump a dead otter on the press table during an Exxon news conference in Valdez. After another chronic technological disaster, Michigan farmers, outraged over the PBB poisoning of their cattle, dumped poisoned cattle on the steps of the capital building (Egginton, 1980: 250). Both the dead otter and the dead cattle were symbols of *unjust* deaths and served as a challenge to the political establishment to do something about the injustices.

Exxon, too, recognized the value of these images and hired a public relations firm specifically to highlight the otter relief centers and their activities on behalf of the wellbeing of otters. Almost overnight, otters became Exxon's poster child, emblematic of the corporation's concern for the restoration of the environment. Some might say that the company successfully co-opted the environmentalists' image and used it toward its own ends.

How people construct the natural world informs us about how they construct and interpret the social world. The deaths of animals were of a magnitude seldom encountered in unnatural disasters. The death of so many animals was deeply disturbing. For those individuals who worked closely with the dead and dying, the experience sometimes was traumatic. Beyond the despair and anguish, however, many people saw the animals as an extension of Alaska's wilderness and what Alaska stood for in their hearts. The death of these animals was symbolic of the *rape* and death of a pristine wilderness—a violation of a revered place. With this remorse, many people experienced guilt; as humans, they felt responsible for the deaths or for the response to the tragedy. When groups attached different symbolic meaning to these animals, there was not only a differential response to the disaster, but also social conflict over the meaning of the disaster.

The rescue and treatment efforts conducted by volunteers, scientists, and Exxon also had an enormous impact beyond the symbolic. Both clinical scientists and volunteers were deeply disturbed by the suffering of the animals, and the experience affected their psyches indelibly. As the experiences of my informants demonstrate, these psychosocial responses to the deaths of *innocents* should be taken into account by disaster researchers. In the case of the *Exxon-Valdez* spill, several people I interviewed seemed disturbed by their experiences with the stricken, dying, and dead animals and reported symptoms closely resembling post-traumatic stress syndrome.

All in all, a staggering number of animals died as a direct result of the spill. Estimates of the number of birds killed range from 260,000 to 580,000. According to one federal study, the best estimate is in the range of 350,000 to 390,000. An estimated 3,500 to 5,500 otters were killed by the spill. Two hundred harbor seals are also thought to have been killed. One hundred and forty-four dead bald eagles were found after the spill; estimates suggest that several times this number were actually killed. Knowing or estimating the number of killer whales that died is even more difficult. One pod was known to have thirty-six individuals. A week after the spill, seven were missing. Within a year, six additional whales were missing. The killer whale pods in the Kachemak Bay area have been seriously depleted. The highly productive and biologically rich intertidal zones were the most severely contaminated. Barnacles, limpets, amphipods, isopods, and marine worms were devastated in heavily oiled areas. Shellfish populations, like mussels, suffered huge setbacks. Eagles and other animals that feed high on the food chain were also vulnerable. Bald eagles were reported to have cached dozens of oil-soaked bird and marine animal corpses in order to feast on the dead (see AOSCH, 1989: 28).

The impact on land mammals was less clear than for birds and aquatic victims. Brown and black bears, minks, river otters, and other mammals frequented the intertidal habitats. Within a short time after the spill, brown bears emerged from their winter hibernation and foraged along the coast for food. Searching the tidal pools for protein-rich kelp, they undoubtedly ingested toxic crude oil. The bears also probably ate the corpses of birds and sea mammals washed ashore. Contaminated with toxic crude oil, the ingestion of these animals would have been fatal. However, deaths of bears are more difficult to estimate, because they were more likely to die in dense forests than on the beaches. Deer, which were also known to feed in the intertidal zone, may have been affected as well. Although the federal coastal surveys reported no mortality attributable to the spill, several informants claimed to have found, or killed, deer whose livers were engorged and smelling of oil. Unfortunately, the oil spill occurred just shortly before spring, the most biologically active time of the year; therefore, it posed considerably more threat to the environment than at any other time. Beyond the statistical calculations of the dead, these animals hold a larger importance. While we have seen that these animals take on a symbolic importance, we also need to recognize the psychological importance they hold.

Many disaster researchers, as well as the public, tend to discount the severity of disasters such as the *Exxon-Valdez* oil spill because human deaths are not involved. Yet, a number of my informants were indignant that Exxon and public officials referred to the disaster as one in which there had been no loss of life, given the tremendous devastation of animal populations. The notion of bereavement has only been considered in terms of the loss of human life (Perry and Lindal, 1970; Krell and Rabkin, 1979; Lifton and Olson, 1976; Titchner and Ross, 1974; Hodgkinson,1989). While I would not argue that the loss of animal life is tantamount in most cases to the loss of human life, nevertheless, the massive loss of animal life resulting from the *Exxon-Valdez* oil spill was important. It figured prominently in many informants' social construction of the spill and seemed to have psychologically disturbed those who worked closely with the animals. Indeed, the effects of this loss mimic, to some degree, the response to a human death. At least one native informant actually did describe the massive loss of life and the destruction of the environment as a greater tragedy than the loss of human life.

I found a significant number of informants who were severely disturbed by this loss of animal life. Some individuals even described their emotional disturbances as a bereavement process. One Alaskan woman writing of her experience of the spill wrote, "I grieve, seeming to take this harder than I can intellectually justify. Caught in the crossfire between head and heart, I am embarrassed at the depth of my pain. I am not alone. Many Alaskans speak of a profound sense of loss, are stricken with the anger, denial and despair of bereavement" (O'Meara, 1989: 19).

Indeed, in at least one community, hundreds of individuals attended a Mass for the deceased animals conducted by religious leaders. Many informants lamented the deaths of the animals, emphasizing that in their eyes the deaths were preventable. Because they resulted from irresponsible acts of men, the deaths seemed especially disturbing. Many individuals might have agreed with the naturalist Joseph Wood Krutch: "[T]he wanton killing of an animal differs from the wanton killing of a human being only in degrees" (cited in Nash, 1989: 76). One old pioneer couple who had lived in Alaska since the late forties stated that they considered "death a natural part of life," but they were disturbed by the "needless deaths" of the animals, whose deaths were "accompanied by great suffering." To them these deaths were "abhorrent" (Linds, August 18, 1990).

Those most affected were the native populations, cleanup workers, and volunteers who collected and identified the animal remains. Workers at the bird morgue in Homer had the undesirable job of sorting, identifying, and counting thousands of bird carcasses. According to one worker, "[E]verybody cried all the time the first month we were working" (Homer Worker, June 12, 1990). These men and women were like the victims of the Buffalo Creek Flood that Lifton and Olson describe: "survivors haunted not just by death but [a] grotesque and unacceptable form of death" (1976: 3). One of the workers tearfully described the situation:

> I'd never seen a disaster like that before. It was really awesome.... They were so coated in oil. They were really a mess.... It would really be a problem whenever you saw a new species. You know, like you got used to murres, 500 or 600 dead murres, but then something new would come in, a little storm petrel or your first puffin. Every time a new species would come in it would get you all over again. Eventually I had to get off dead birds. I had this nightmare about them and I woke up in the night crying and I started digging around in my blankets because I could hear a little bird down there. Yeah, it was a horrible experience. To this day I can't deal with people who down play the spill. (Bird Morgue Worker, August 9, 1990)

While some natives were moved by the suffering and deaths of animals, some were moved in ways different from the nonnatives. Many of the natives who were accustomed to witnessing the death of animals, and inflicting death in the course of their subsistence activities, were, nonetheless, deeply shocked and saddened by the animal deaths. Certainly, the natives were disturbed by these losses because their own survival depended in large part on the abundance of wildlife resources. Nevertheless, they were also clearly shocked by the suffering of the animals themselves. For example, while the death of otters was of great symbolic importance to the media and the general public, the deaths of these animals held a very different meaning for some native people. Some natives do not see this animal in the anthropomorphic terms of many nonnatives; rather they view the otter as a pest who competes with them for shellfish and crab.

This view is not unlike many nonnative fishermen who also see otters as a competitor (Oldham, August 17, 1990; Sheppard, August 17,

1990). Nevertheless, the deaths of these animals upset many of them. One native beach cleanup worker, formerly an otter hunter himself, described his reaction to these deaths. As he hugged the crude oil-soaked corpse of one otter, he said to his coworkers, "These little fuckers didn't deserve to die like this."

Some natives mourned not only the loss of one species, but the loss of all marine mammals, birds, and fish (P. Totemoff, April 11, 1990; Hedrick, April 18, 1990; Kompkoff, April 18, 1990; Kvasnikoff, April 23, 1990). One native compared the *Exxon-Valdez* disaster with the earthquake and tidal wave that destroyed his village and one-third of its inhabitants twenty-five years earlier. Although the natural disaster had resulted in considerable loss of human life, including members of his own family, he felt that the loss of animal life and the over-all effect on the environment were even greater from the oil spill. For him the loss of so many different animals and the almost total destruction of the natural environment surrounding his village were far worse than the loss of human life alone (J. Totemoff, April 11, 1990).

Whether they were hardened fishermen or college-aged environmentalists, many informants frequently broke down sobbing when discussing the degree and scope of animal deaths they observed. On any one occasion, employees and volunteers from the cleanup crews and animal rescue patrols would discover anywhere from one or two dead animals to several hundred. One month after the spill, in a single day on a single strip of beach one-half mile wide, workers found six dead sea otters, three dead land otters, three dead eagles, and hundreds of dead birds (Homer News, April 27, 1989). The animal corpses, like human corpses from a natural or man-made disaster, were often severely deformed and transmogrified by the crude oil. This transformation was so total that the rescue workers had to devise a beak identification chart in order to identify the species of birds. These transformations had an extremely disturbing effect on the individuals who had to deal with these corpses on a regular basis.

The horror of gathering animal corpses from the beach is best illustrated by the story of Gore Beach. Located on the outer coast of the Kenai Peninsula, the beach was heavily oiled. When the rescue crew arrived, they found the beach shrouded in a dark mist. In a half-mile swath of beach, they found crude oil that was shin-deep. There, among the oil, they found almost three thousand bird corpses representing a variety of species. Elizabeth Wolf of Homer, a member of the Gore Beach crew, gave the following account:

> I stepped out on a devastated Gore Beach. In front of me lay not hundreds, but approximately 3,000 dead birds mixed in six inches to [twenty] inches of crude oil. It coated the entire beach and floated in the water. My assignment of bird rescue was quickly transformed into bird collection. Knee deep in oil became elbow deep as I plunged arms into the thick substance to pull unidentifiable birds from a gripping grave. Our crew soon collected 900 corpses, 631 in one pile alone. This was my first introduction to the quiet side of the Alaska oil industry, a first impression that will not fade. (NWFH, 1989: 151)

Anxious to remove the poisoned bodies from the food chain but unable to remove them safely from the beaches, the crew foreman decided to burn the corpses. In order to document an event that might later be used as evidence against Exxon in court, the foreman had the procedure videotaped. Numerous copies of the tape circulate in the homes of Homer residents. Slowly the bird corpses are laid out on the beach. Identification of species is almost impossible. Many of them are large birds. The camera pans the length of the beach and shows the viewers the neat rows of animal bodies. Like watching some exotic mortuary rite, workers then build a funeral pyre seven feet high and begin tossing the heavy bodies on top of the wood. Someone lights the wood, and flames soar as workers continue to throw body after body into the fire. The flames' light and shadows flicker over the workers. Some throw the corpses onto the pyre; others stare silently. The fire roars as it consumes the oil-soaked bodies. This is no mythic ritual however, and the birds do not rise like a phoenix from the ashes. The devastation is complete. This event became one of the focal images for my informants and other Alaskans. Several of my Homer-area informants and one native informant from Port Graham cited this videotape and the incident while discussing their reaction to the animal deaths.

Those who suffered the most from this exposure to animal suffering and death were the people who either cared for the injured or identified and autopsied the remains. Numerous people of all ages and backgrounds worked with the otters at the recovery centers in Valdez and Seward. They recounted to me the blood-curdling screams of the agonizing otters and the seemingly painful spasms of their bodies in the throes of death. One woman said that watching the suffering of the animals was "like watching your best friends being tortured to death" (Volunteer, July 3, 1990). Not one of these informants could, even more than a year later, describe this experience without breaking down into tears. Likewise, those volunteers who collected, tagged, and identified the oil-soaked bodies of the birds recounted horrible, recurring nightmares as the result of dealing with death day after day. Even the seasoned biologists, used to clinical dissection, were deeply disturbed by this task. Dr. Calvin Lensink, who had worked for the U.S. Fish and Wildlife Service for thirty-three years, said, "It was terribly depressing seeing all these birds. There is a feeling of insult when a bird comes in sopping with oil. Unrecognizable" (Davidson, 1990: 148).

The Impact of Animal Death on Humans

Bereavement cannot be dismissed as a mere metaphor for the informants' experiences. The symptoms that many informants exhibited require us to examine closely the serious consequences of these massive animal deaths. Almost all individuals who worked directly with the dead and dying animals, and many who simply witnessed the destruction of wildlife, told me that they suffered from recurrent nightmares from their experiences—nightmares that some still struggle to repress. Numerous workers told me they suffered from loss of sleep, loss of appetite, and depression during the period that they worked with the animal corpses. Some struggled with symptoms for weeks. Therapists in Anchorage led grief workshops, which were attended by a number of Homer area citizens. Several individuals stated that they quit paid clinic and cleanup positions when they felt their own mental health was in jeopardy. One woman told me that she struggled with depression, which she attributed to her volunteer work with the animals, for almost a year.

The fact that so many informants continued to suffer from nightmares and depression indicates the importance of this dimension of the spill. We should reexamine the notion of bereavement in light of this finding and explore its significance when discussing the impact and severity of man-made disasters. How the death of animals affects humans is not really clear. However, given the evidence from Homer residents and the resemblance between the psychological response to animal deaths and the bereavement process for human deaths, this phenomenon clearly deserves further attention by chronic technological disaster researchers and bereavement researchers. There is reason to believe the deaths of animals adversely affected some of the rescuers.

Other technological disasters have caused animal deaths as well. However, most of these deaths were domesticated animals, and their toll was nowhere near as high as in the case of the *Exxon-Valdez* oil spill. Domesticated animals died as a result of the chemical accident in Bhopal, the nuclear accident at Chernobyl, and the PBB contamination in Michigan. Many Michigan farmers were forced to kill their contaminated livestock. Egginton (1980) reports that many farmers were unable to talk about slaughtering their cows without emotion and tears in their voice. Dairymen and their families often wept as they shot their cows. One dairy farmer that I spoke with several years after the incident told me that the loss of his herd had a tremendous emotional impact on his life. One Alaskan Native woman in the village of Tatitlek discussed the death of the animals from the *Exxon-Valdez* oil spill as being like "having some sick invader rob your farm and slowly slaughter your pets and domesticated animals one by one as you were forced to watch" (June 8, 1990).

A number of researchers (Raphael, 1991; Raphael, 1986; Raphael, et al., 1983–1984.; Duckworth, 1986; Hodgkinson, 1986) have observed that rescue workers who confront scenes of physical revulsion often suffer from adverse mental health effects. For example, one of the earliest studies of this kind was in Sydney, Australia (Raphael, 1991: 1346). Following a rail crash, 70 percent of the rescue workers suffered "transient symptoms of post traumatic distress: nightmares, anxiety, flashbacks." While this research pertains to the loss of human life, the parallels to animal rescue workers is notable. These encounters with death seem to have similar impact on the rescue workers. Like rescue workers in disasters where human life has been lost, many of these workers experienced irritability, loss of interest or withdrawal, and disrupted interpersonal relationships. In this case, it is important to remember that the stress of rescue workers was compounded by the fact that they were both rescue workers and victims of the disaster itself.

Social Conflict

In our analysis, we have to move beyond an examination of the symbolic and the psychological consequences into the area of social conflict. Conflict and uncertainty arose between both native and nonnative people as well as between scientists and animal rights activists. For instance, many individuals in the Fish and Wildlife Service told me they were against the rescue mission and only reluctantly participated. The controversies that ensued underscore the tension that exists in many man-made disasters between scientists and laymen. Moreover, the differing responses between native and nonnative people also serve to foreground the cross-cultural impact and different cultural interpretations of the disaster. For some native peoples, the animals that were killed by the spill were not anthropomorphic creatures and romantic embodiments of wildlife. They were an important food source. Some nonnative people did not see the otters in sentimental terms either. Nonnative old-timers and some fishermen spoke of individuals concerned for the otters as "otter-huggers" (Kilcher, August, 1990). The devastation of some species in their subsistence grounds seriously threatened their way of life.

Some animal rights activists accord animals the status of children. Like children, they are considered innocent and helpless and need humans to protect them especially from human abuse. Interestingly enough, a great many people stated that the two groups that suffered the most from the spill's effects were animals and children. Many people stated that this was the ultimate tragedy. Since both groups are innocent, the responsibility of good citizens is to protect them from harm. Many regretted their inability to perform this role during the spill. Others used this abuse as an example of the callous disregard large corporations have for the sanctity of life.

When I asked one long-time resident of the Homer area why the suffering of the animals was so disturbing, she replied:

> Well they are so innocent, they have no blame, no guilt. They don't get anything from the development, from oil. They just get grief. ... They couldn't know what was happening to them. They were just suffering. Extreme suffering. ... [P]eople in Alaska, myself included, tend to view Alaska as an entity. And the birds and otters are an extension of that. ... [W]e feel good when we see them. ... And then to see them suffer and not know what happened. I guess partly the anger and suffering there for us is the guilt that we allowed it to happen. And it was our responsibility to try and keep the world good for them. And we didn't meet our responsibility. ... I suppose there is another reason why people wanted to get out there and do something. Pick up the animals and try to help get over the guilt ... assuage the guilt, overcome it. It was wrong then, but I am doing something now. ... [I]t just seems totally unfair that the animals should suffer for our greed. (O'Meara, June 14, 1990)

The idea that animals are innocent, like children, was central to many of the informants with whom I spoke. According to their way of thinking, humans have a responsibility toward animals. Many felt that as humans, we had neglected and violated that responsibility because of the spill. Alaskans were not alone in expressing this view. Many of the volunteers from around the nation who worked at the animal centers spoke in similar terms. A good many of these people said they came to Alaska to help the animals and do *penance* for mankind.

A number of people that might be loosely labeled animal rightists likened native people to animals because they were perceived as innocent and suffered needlessly because of the spill. Harvey Feit (1991) observes this tendency as well: "For Euro-Americans, not only do animals become permanent children, but so do indigenous peoples, also 'children of nature' become so associated. Indigenous people are seen apart from the world of Man and are viewed as a part of Nature. Thus, they are more innocent than most humans and require our guardianship." This perception of the indigenous people as children of nature is understandably objectionable.

Some of the natives had vastly different feelings about the death of the sea otters. For generations they have hunted the otters and are now, because of their native status, the only people allowed to shoot them. Many natives greatly resented the enormous amounts of money Exxon spent rescuing and rehabilitating these animals. Exxon spent an estimated $80,000 on each rehabilitated otter (*New York Times,* December 31, 1991). Exxon provided about $18.5 million to capture and treat 357 otters (*New York Times,* October 21, 1990, C4). By the time the project came to a close in August 1989, 222 otters had been collected, but only 18 percent survived the spill and rehabilitation process.

Some native people told me that they thought such an investment demonstrated that Exxon placed the value of and concern for native people below that of animals. For example, in Prince William Sound, two villages received a barge load of food from Exxon in partial compensation for the loss of the villages' subsistence grounds. Natives in both villages became sick, and some were airlifted to Anchorage because of food poisoning. The barges contained seafood that, it was later learned, was actually shrimp and crab meat intended for aquarium food, obviously unfit for human consumption. Some natives contended that the shipping of the food to their villages was not a mix-up, as claimed by Exxon, but a deliberate disregard for the value of their lives. One woman in Tatitlek told me that this incident illustrated how, "Exxon considered the natives no better than animals" (June 13, 1990). After the spill and during the cleanup project, many native villages had a difficult time obtaining funds from Exxon that would address the adverse social consequences of the spill. One native leader told me he thought it was obscene that Exxon would spend so much on animals and so little on people (G. Kompkoff, April 18, 1990).

Natives, however, were not the only people critical of the rehabilitation process. Many scientists and lay people have criticized the tremendous cost of the project, and many have argued that little scientific information was gained in the process. One such critic is Dr. James Estes who has stated, "[T]here is a general feeling that a lot of otters were saved, but very little was accomplished, despite all the money that was spent" (*New York Times*, December 31, 1990). In the journal *Science*, Estes stated, "[P]lanning [based on studies of released otters from the *Exxon-Valdez* oil spill] of this kind tends to lull public and policy makers into a false sense of readiness" (December 31, 1990). Estes questioned whether the wildlife population was threatened by the loss of some individuals, even if that loss may number in the thousands, as in the case of the sea otters in the *Exxon-Valdez* tragedy. He questioned the amount of time and money invested to save, ultimately, relatively few individuals.

Estes was not alone in his criticism. Many in the Homer area were also critical of the rescue project. One fisherman alleged that the project was mostly a publicity stunt for Exxon. While this man was sympathetic to some of his friends who became involved in the project as volunteers, he told them he thought, "they were being taken advantage of by Exxon." Another fisherman had more to say about both the otter and bird rescue projects:

Anthropomorphism is one of my pet peeves in life. I know people donate thousands to save a baby seal, but the threatened Costa Rica Iguana gets little or nothing. I try to teach my little girl to treat all animals with respect, not just the animals that humans feel close to. Spending ridiculous amounts of money to achieve so little is absurd. Some of these bird boats were fifty-footers and they were paying them $5,000 a day for the use of them. These things are going out and burning up 500 gallons of fossil fuels a day, and by summer they were down to finding maybe three birds a day. I mean that made no sense. I don't think it makes sense to spend 500 gallons of fuel to find one bird. I realize they need the bodies for litigation, but by that time they already had 35,000 dead birds! .... I don't feel any closer to otters than to sea lions or iguanas. I mean they are nice animals to have around, but spending $80,000 a piece on one? I mean the money could have been better spent going into a research center on animal life in Kachemak Bay. (Choate, September 12, 1990)

Many associated with the project disagree, maintaining that the project was worthwhile. Dr. Randal Davis, the biologist who ran the project for Exxon, is one of the project's defenders. Davis stated, "[W]e learned an incredible amount about how to do it better in the future. What we learned was worth every cent" (Dold, October 21, 1990, C4). Another defender is Nancy Hillstrand, a biologist who directed the otter center in the Homer region. She believes one should spend whatever it costs to save an animal: "Those animals were in a lot of pain, gouging out their eyes, chewing their paws off. The otter losses wouldn't cause a problem in the population but I believe in the individual animal. When you are listening to an otter scream, you can't look into its eyes and say, there's a lot of you around, so I am not going to do anything. I did whatever I had to do" (*New York Times*, December 31, 1990).

The different attitudes expressed in these statements by Hillstrand, Estes, and others are significant not only for what they tell us about the worth of the project, but also for what they reveal about their different views about animals. The importance of this difference will be further discussed below.

A number of nonnative informants and one native informant told me that while otters were being released back into the wild in the Kachemak Bay area, a couple of native hunters sat off shore in their boat and shot several of the released animals. There was some speculation that the resentment over Exxon's lavish treatment of the otters spurred the shootings. One Homer-area resident, who considers herself an animal rights activist, was appalled by this behavior and stated that the incident indicated the "cruel attitudes natives have toward animals" (June 12, 1990).

While native and nonnative people held very different views about the deaths of the animals, and seemed to process different worldviews about the ontological status of animals in general, there also emerged radically different paradigms between nonnatives who worked at the animal treatment in the Homer region. Wildlife biologists, veterinarians, and people with science degrees held a very different perspective about what they were doing than laymen, among whom many were declared animal rights activists. Scientists, in general, tended to be most concerned with threats to an entire population or species. Animal rights activists, who served as both employees and volunteers in the rescue efforts, tended to be most concerned about the suffering of individual animals. At the Homer otter center, this disjunction created a terrible rift between coworkers, which eventually spread to the town itself. At the center of the controversy lay the issue of the release of rehabilitated otters into the wild and a debate over the way in which the oil spill's tragic effects for animals might be used to gain further knowledge about these animal populations. In examining this division between the two groups of nonnative Alaskans, we can arrive at an understanding of some of the differences in worldviews, values, and social conflicts between the two cultures.

Significantly, the conflict developed primarily at a center where 122 otters were being treated. The employees at the bird treatment center remained united. It is important for understanding the conflict that did develop to spend a moment analyzing the reasons why a similar conflict did not occur at the bird center. For one thing, animal rights activists tended to be drawn to the otter treatment programs. One layperson familiar with both groups of people had her own beliefs about the differences between workers at the bird center and those at the otter center: "[T]here is just something about people attracted to otters that just makes them a voodoo element. ... People working at the bird center were not as emotional as the otter workers" (June 8, 1990). Further, the animal rights contingent tended to emphasize the individual animal, claiming that each animal has a life equal in moral terms to a human life. As one individual described the difference:

> Trained as scientists, the Fish and Wildlife people are really concerned that a species survive. They are not concerned that an individual animal survive. Individual animals don't survive all the time. The animal rights activists are very concerned with the welfare of an individual animal, even if the species is going down the tubes. That just goes against the scientific background of the Fish and Wildlife people. They are concerned with knowing certain things about where the population is going and what needs to be done to insure the population remains healthy. (Benson, June 31, 1990)

This belief in the individual involves a certain anthropomorphization of the animal individual, and it is easier to anthropomorphize otters than birds. Otters not only have large neotenous eyes and appear very doglike in their behavior, but they also have a social organization that more closely resembles monkeys or humans than that of birds or other affected animals. For these reasons, the otter center proved more fertile ground for conflict between the worldview/values of scientists and those of the animal rights activists. To this controversy, we now turn our attention.

This dichotomy was expressed earlier in the statements made by Estes and Hillstrand about the expense of the otter rehabilitation project. Estes, the scientist, in his criticism of the endeavor, was not concerned with the loss of individual life, but with whether the species was threatened. Hillstrand, the director of the Homer otter center, stated quite explicitly that she was concerned with the individual animal. Contained within this statement is another important distinction between the worldviews of some members of these two groups. The emphasis on the individual of the species held by some animal rights people reflects their projection of American cultural assumptions about the primacy of the individual over society. The scientific perspective held by some scientists is more concerned with social survival.

With regard to these differing paradigms, views on death can differ, too. Many animal rights activists are deeply disturbed by the death of even one animal. The death of individual animals, even many individuals, does not disturb many scientists as much as a threat to the ecosystem. Many scientists accept death as a natural part of the life cycle. They are not as deeply disturbed by the death of an animal or even several animals as they are concerned about the survival of the species (interview with Poppy Benson, wildlife biologist). At least this model is the stereotype that is widely shared, even by many scientific informants. However, we have to differentiate between what scientists say and what they actually do. Even some of the examples of scientists' responses to the death of animals included in this chapter do not strictly adhere to this model. The following quote reveals the empathy of one scientist for the individual, even as she principally focuses on larger concerns. This Ph.D. ecologist was watching a murre struggle in oil-slickened waters:

> As I stand on the bow, a deep grief begins to emerge in me. Beginning in my belly, it rises through my being and surfaces as sobs which make my soul shake. The enormity of the horror washes over me, watching this lone bird beginning to die, struggling to live, in the stench of gasoline. The ecologist in me has known for days the destruction the oil is causing [and the] dramatic changes in the entire ecosystem. Not just the death of individual animals, plants, plankton and lifestyles; but the total changes in the energy flow through the system.

The dynamic flow of energy, nutrients and life is altered, blocked. The functioning chain of life, ebbing and flowing from one organism to another, from one generation to the next, no longer functions. (Spencer, 1990: 29)

According to at least one informant, biologists were also constrained by their professional image:

These biologists were struggling to take an objective point of view, pulling dead otters and dead birds out of the sea. And they're supposed to be scientists. It completely changes their view of what they are trying to do. I had a biologist break down and cry in front of me—a friend of mine who had been working on the spill and was out there since Day One. It was a tremendous emotional thing for him that he had to suppress, because he was a biologist working on the spill. I had another biologist break down and cry. She said to me, "I can't show you my emotions about all this. As a scientist, you lose your credibility if you show any emotions about what you are doing." (Benson, June 31, 1990)

Indeed, Page Spencer, the ecologist cited above, came under considerable criticism from some people I met because, as a scientist, she was considered unobjective and too emotional. Some informants considered her story unprofessional and embarrassingly emotional. However, another informant, who was a scientist herself, felt that Spencer did the profession a service by speaking the unspeakable.

Immediately after the spill, a makeshift otter treatment center was set up in Valdez, and Exxon eventually contributed money to its support. A number of Homer residents became involved with otter recovery by volunteering at this center. When the center expanded, some of these people were placed on the payroll of the center, while others remained at the spill as volunteers. As the oil spill flowed beyond Prince William Sound and as the number of captured otters grew, another center was established in the city of Seward. Both facilities had drawbacks in that they were, as one Homer man expressed it, "too industrial." For example, the center in Valdez was located in a remodeled community college gym. Large industrial fans overhead buzzed nonstop, and scores of people moved about constantly. The atmosphere was not the most salubrious for otters used to the isolation of the wilderness. Increasingly, rescue workers recognized the need to establish a center that was more secluded and peaceful, one that could serve as a way station before the otters were released back into the wilderness.

Also during the month of May, the spill's impact on the Gulf of Alaska and Kachemak Bay created a need for a center in Homer. Nancy Hillstrand, a charismatic and resourceful person who both lived in the Kachemak Bay area and had worked for the state fish and game department, did what many socially conscious Homer residents did during the spill's crisis period: she decided to take matters into her own hands and organize a center. The center in Valdez and the one in Seward were established, in part, by Exxon involvement in the process. Exxon and the people affiliated with the otter center in Valdez were hoping to open another center. Hillstrand, who contributed to the start of a volunteer center in Homer, eventually teamed up with these people and obtained some Exxon support to fund the costly endeavor. Three or four of the Homer residents who had worked at the center in Valdez returned to help Hillstrand establish the new facility.

A scenic, secluded spot at Jackaloff Bay, a cove across the mouth of Kachemak Bay from Homer, was chosen, and outdoor water pens were built. Eventually, almost one hundred workers, paid and volunteer, participated in the labor-intensive process of caring for ninety-six otters, the greatest number ever held together in captivity.

Many informants recounted the excitement and thrill of the early days of the center. They felt privileged to be able to observe the otters, perhaps even occasionally hold one. The setting of the site was spectacular. For many it provided a haven from growing strife and tension in the city of Homer. Life at the center appeared to be idyllic (according to statements of several volunteers). For those employed, the pay was beyond the imagination of most of the employees. The conditions of work, however, were stressful, with long work shifts. The site, however scenic, was remote and isolated. For a long time, the center had no phone, and travel back and forth across the bay was problematic. As people became increasingly isolated from their family and friends, tensions grew and stresses increased. Among the many single people employed there, romances and then jealousies flowered. Some couples, separated by great distances, broke up. A few employees got divorced. The person in charge of husbandry later told me his primary concern became avoiding the transference of this tension to the otters themselves (Interview with Kurt Marquardt, August 11, 1990).

By mid-summer these frustrations and tensions channeled into a controversy that would result in severe divisions. Uncertainty about the best ways to treat the otters lapsed into rigidly opposing views. As one participant said, "It became so vicious that friends of employees there, who never worked there, took sides and stopped talking with people on opposing sides" (Marquardt, August 11, 1990). A schism developed between the scientists and the animal rights people over a number of issues. The biologists at Fish and Wildlife became centrally involved by issuing directives from afar. Center employees sympathetic to government policies turned out to be people who had scientific backgrounds. Former vet's assistants and people with master's degrees in animal husbandry found themselves in conflict with the animal activists who were in the majority.

Several issues emerged that highlighted the underlying tension between the two camps. First, the Alaska Fish and Game Department became quite concerned when it learned that some of the otters had a herpes infection. They were worried that if the animals were reintroduced into the wild, they might introduce an infection acquired during captivity. Foremost in the minds of the Fish and Game people was an epidemic among seals that occurred in the North Sea some years before, when the seal population was all but obliterated because a former captive transmitted canine distemper to the population. Wildlife biologists wanted to delay the release time in order to investigate these health risks. They also wanted to secure blood samples to screen for contagious diseases and to decide which individuals were healthy enough for the implementation of radio transmitters.

This latter issue became a central point of contention between the scientific and the animal rights factions. The animal rights contingent did not want the otters to be handled or disturbed by the taking of blood samples. They strongly objected to the use of flipper tags and adamantly opposed the implementation of radio transmitters. The radio transmitters were small, cigarette-size radios that would be surgically implanted in the back of the otter's necks. From the animal rights perspective, these procedures were immoral. They urged immediate release. The U. S. Fish and Wildlife people, who had allowed the center to operate fairly autonomously, intervened, insisting on the procedures (Benson, July 31, 1990). Hillstrand, the center's director, strongly opposed the procedures and was upset by the intervention and her loss of authority and control (Losbaugh, August 2, 1990).

As the dispute wore on, many activists in the Homer region, who had friends on both sides of the dispute, became concerned that the animal rights people were playing into the hands of Exxon. While they did not question the motivations of the animal rights people, they were concerned that this disruption would ultimately help Exxon. The animal rights advocates were interfering with the collection of data on the long-term effects of oil on sea otters. Without this knowledge, many people thought the legal claims against Exxon would be undermined (Benson, July 31, 1990). Others alleged that Exxon was doing everything it could to insure that long-term effects of the oil spill remained unknown. The scientists wanted the radio transmitters to track the otters and plot their survival rate.

Nevertheless, their plan was alleged to be ill-conceived, because they did not have a good control group, and they could not account for many intervening factors (Christianson, August 11, 1990). Hillstrand herself criticized the research design. However, she and her supporters were not interested in proposing a more adequate research design. The Fish and Wildlife plans were deemed immoral and were summarily dismissed. Emotions at the center became charged (Losbaugh, August 3, 1990; Christianson, August 11, 1990; Baugher, August 9, 1990). Some workers proposed cutting the nets of the otters and releasing them in the dark of night. In Valdez, some members of the rehabilitation team cut the nets of the floating pens and allowed thirteen of the sixty-five otters to escape before the surgical team arrived to implant collars. Five of these otters were later recaptured (Marquardt, August 11, 1990).

The Homer situation became critical when biologists and veterinarians began implanting the radio transmitters. After the first of these surgical procedures, one of the otters, which had been placed back in his watery pen, lost consciousness and drowned. The animal rights activists became incensed, insisting that this was a clear example of why they wanted no more interference with the otters (interview with Kurt Marquardt, August 11, 1990; Tina Baugher, August 9, 1990). Supporters of Fish and Wildlife made a point of stating that the otter center staff was so disgruntled about the procedure taking place that they, in effect, walked off the job and, as a result, the otters lacked proper supervision. These supporters insisted that the otter would not have died had the staff properly monitored the pens. One of the staff people physically attacked one of the veterinarians (interview with Shana Losbaugh, August 3, 1990; Poppy Benson, July 31, 1990).

Events deteriorated even further. Apparently, files were destroyed. Many people said that Hillstrand, who was described by all as charismatic and deemed by many to be willful, falsified the animals' records in order to make it appear that they had already had blood samples drawn. When one of the workers who was in charge of the records (and who was unaware of Hillstrand's changes) unknowingly disclosed the discrepancy, she was reportedly fired by Hillstrand, allegedly because, according to other workers, the worker made Hillstrand look like a fool. In fairness to Hillstrand, none of these allegations were ever proven. People on both sides alleged that Hillstrand and some sympathizers contemplated releasing all the otters on their own initiative before the biologists could intervene.

Eventually, they were said to have dismissed the plan, because they realized that doing so would overload the environment and stress the Kachemak Bay otter population. Unable to afford costly helicopter or boat transfer of the otters to alternative locations, they dismissed the idea. Accusations and rumors about Hillstrand abounded. She became a controversial person to some and a hero to others. Some people praised her because they thought she placed the welfare of the animals above the concern of science. Many of her detractors credited her with vision, dedication, and resourcefulness but claimed she became obsessed with control, self-righteous and determined to have her way at all costs. Many detractors were also empathetic, saying that Hillstrand simply was a victim of the stress surrounding the spill activities. However accurate these assessments may be, they tend to personalize the issues and overlook the real reasons why the split occurred, ignoring the fundamental philosophical differences between the scientists and the animal rights people.

This schism, while unique in its focus on animal rights, characterizes the tension and mistrust that commonly occurs between scientific experts and laypeople in the wake of a chronic, technological disaster. While the tension that emerged around this issue was certainly the unique consequence of a massive disturbance, it also underscores the mistrust that many disaster victims have of experts. Because chronic technological disasters are man-made, their victims often believe that the disaster would not have occurred if these scientific experts had known what they were doing in the first place. All scientists and experts are usually conflated and held accountable. A loss of trust in experts is one of the outcomes of man-made disasters. This mistrust of scientific expertise followed in the wake of chronic technological disasters in Love Canal, New York (Levine, 1982), Seveso, Italy (Reich, 1991), and Woburn, Massachusetts (DiPerna, 1985).

In our discussion of social conflict, we have seen the effect of different worldviews and values on various responses to the suffering and death of animals. Less easily discerned, however, are the ways in which each group manipulated symbols and issues in accordance with this own self-interest. I earlier suggested that Exxon used the photographs of animals to underscore their concern for the environment, turning animal suffering into a public relations boon through their well-publicized funding of rescue efforts. The self-interest motives of other groups are less apparent. These views were also forged by each group's political or material interests. Thus, while some of the natives do have a very deep reverence for animal life, very different from that of nonnatives, the native groups also employed these differences in their public discourse toward their own political ends. In this rhetoric, their reverence for all life was juxtaposed to Exxon's seeming indifference to the suffering of both natives and animals. This stance could only strengthen their class-action suits against Exxon. At the same time as they denounced the inhumanity of capitalism (i.e., Exxon), they also harvested otter pelts for their own financial gain.

Moreover, nonnatives also took positions consistent with their own self-interests. The position of animal rights activists were laudable in many ways. However, their paid employment and social status often depended on the continuation of rescue efforts, even as costs of the effort began to outweigh gains for the animal populations. A close friend to many of the activists, one Homer fisherman opined that the animal rights people "feared the severance of their highly lucrative paychecks with the termination of the projects" (June 7, 1990). It must be pointed out that this was highly speculative on his part and that he had no information to substantiate his claim. Nevertheless, this dependence on Exxon money turned out to be ironic, since many people associated with these projects had started out as unpaid volunteers, often without any material motivation in mind. Nor did the scientists act without potential self-interest. Undoubtedly, at least a few scientists were thinking of their own careers and publications when they argued for the collection of additional data on animal behavior. Therefore, probably neither side in the otter rescue or radio collar controversies was purely value-driven.

In pointing out possible ulterior motives, I do not mean to dismiss the significance or sincerity of the different worldviews or the controversies and uncertainties involved. However, at the same time, that these values are not formed in a vacuum must be acknowledged. Material self-interest always plays some role in the decisions taken by any individual or group. In any social situation as deeply charged as the responses to animal deaths following the *Exxon-Valdez* oil spill, individuals will find the political and material realities of their actions both shaped and reshaped by their values and worldviews.

The animals who suffered and died as a result of this man-made environmental disaster took on a special symbolic importance for Exxon, Alaskans, and Americans at large. Yet the creation and management of these symbols depended both upon preexisting ideas and the relation between humans and animals and upon the depth of the psychological impact of this aspect of the tragedy. Symbols and psychology intertwined, with each constructing and reconstructing the other. Which symbols were chosen, how they were manipulated by various groups, and what the results of these clashes would be, form the social response to this disaster. Differing worldviews set the stage for a clash of values and for the resultant social conflict. The responses of various groups cannot be understood without seeing them in the light of their preexisting worldviews. The rumors, gossip, and moral judgments that emerged around these incidents reflect the morally charged climate that loomed during the cleanup period and still lingers today. In short, none of the positions taken by all sides was value-neutral, a condition that characterizes all controversies and uncertainties in the wake of disaster.

3. "What You Don't Know Can't Hurt You"

On a stormy day in February 1993, in Dunross Parish, adjacent to the tiny village of Scatness, an elderly man entered his small, modest cottage on the oil-stained moors of the Shetland Islands. The man wiped his hair slick with the palm of his hand as he pulled out a chair in front of the peat-burning hearth and motioned me to sit down. Except for his twenty-five years in the merchant marines, eighty-year-old Colin Andrews has lived on the tip of this island most of his life. Andrews can trace his family's history here back ten generations.

> "I am too old to worry about myself," he began as he took a sip of bourbon from a bottle at his feet, "but I am worried about my son and my grandchildren, who live a couple of miles up the island on their croft where they grow a small truck of garden vegetables and raise a few dairy cows and a herd of sheep. One of their cows is ill and four of the sheep were found dead on the moor. It's bad enough that the crofters are losing some of their livestock and have lost all of their crops but many of us are also worried about people's health as well. My daughter in-law is five months pregnant, and my son's three boys have to stand outside twice a day to wait for the school bus. The people in Lerwick assure us that there is no risk to the public's health but, like many, I have a hard time believing that. I wish they would give us more information. There is just too much we don't know and too much they are not telling us." (Andrews, February 12, 1993)

On January 5, 1993, the American-owned oil tanker *Braer* went aground at Garth's Nest on the southern tip of the Shetland Islands. Amid

The oil tanker Braer went aground at Garth's Nest on the southern tip of the Shetland Islands.
Photograph © Chris Gomersall.

a hurricane, the tanker was smashed into several sections and spilled 84,413 tons of Norwegian crude oil into the North Sea. This spillage was approximately twice the amount of the oil spilled in Alaska, by official estimate, following the wreck of the *Exxon-Valdez*. Coincidentally, this wreck, the largest since the catastrophe in Alaska, also occurred at 60 degrees north latitude. Within a few hours, an oily fog, 200 feet high, moved over the southern end of the island.

The Shetlands are an island archipelago located approximately 250 miles due west of Bergen, Norway, and over 250 miles north of Aberdeen, Scotland. As in Alaska, many of the 22,500 inhabitants depend on maritime resources for their livelihoods. It is a maritime community with rich fishing grounds. In 1993, there were approximately 500 fishermen on 100 fishing boats, sixty-three fish farms, and twenty fish factories. Almost 30 percent of the Shetland workforce was employed in the seafood industry (Moncrief, 1993). Until the 1970s, economic livelihood in this remote region centered largely around fishing, wool, crofting, knitting, and subsistence farming. After the discovery of large quantities of oil sixty miles off Shetland's eastern shores, Sullom Voe, the largest oil port in Europe, was constructed on the northern tip of the island (Rosen and Vorhees-Rosen, 1978).

While the two spills are significantly different, the Shetland Islands oil spill resembles the spill in Alaska in some important ways. In Alaska, a minority ethnic group, Native Alaskans, were most adversely affected by the disaster and the cleanup that followed. In a similarly disturbing way, the native people of the Shetlands were also most adversely affected by the spill and the response effort. In both instances, the way of life of marginalized groups was threatened.

Other glaring similarities are obvious, such as disputes over the use of dispersants, concerns about public health, the health and safety of workers, concerns about the contamination of the food chain, and controversies over the withholding of information. These concerns generated considerable anxiety among the residents.

Sensing a story even bigger than the *Exxon-Valdez* spill, 600 reporters from all over the world flocked to this remote northern island within hours. Members of the press corps were not the only people who had the *Exxon-Valdez* on their mind. Shetland Island officials and the Scottish Office officials feared a tragedy of the magnitude of the Alaskan disaster and expected criticism similar to the controversies surrounding *Exxon-Valdez* clean-up effort. The ferocious winds blew not just oil vapors, but

73

oil droplets inland, clear across the island, coating the houses and land of nearby communities with a brown, viscous slime. Windows in the houses of Hestingott, Scatness, and Quendale were smeared with oil. Despite hurricane-force gale winds and rough seas, officials decided to spray over 120 tons of chemical dispersants within forty-eight hours.

Malcolm Green, executive director of the Shetland Islands Council, demanded that the U. K.'s Marine Pollution Control Unit (MPCU) assure him that the dispersants had been tested for safe use around humans. The MPCU assured Green the dispersants were safe. Later he told me he was furious when he learned this statement was less than accurate. Ever since the sinking of the *Torrey Canyon* oil tanker off the coast of Great Britain in 1967, dispersants have had a bad reputation. The British dumped a phenomenal 10,000 tons of dispersants on 14,000 tons of crude oil. When the dispersants failed to work, the Royal Navy Air Corps bombed the tanker, attempting to set it afire. By then, the oil had mixed with the water and the attempt failed.

Dispersants are solvents used to break down the cohesiveness of the oil. They are sprayed on an oil spill in order to remove the oil from the surface of the water. Chemical agents cause the oil to enter the water column in tiny droplets and become diluted, ostensibly until they reach such low concentrations that they are harmless to the environment. Early dispersants were considered toxic, but current ones are believed somewhat less toxic (National Research Council, 1989). In rough seas and high winds, such as the conditions surrounding the wreck of the *Braer*, dispersants have been found to be ineffective. While virtually ineffectual on the crude oil along the rocky coast, the dispersants contaminated the Shetland Island's prime agricultural land and most of the nearby inhabitants, as well as several thousand sheep and cattle that had no shelter from either the storm or the chemical sprays.

Environmental Health Issues

Four communities lie in close proximity to the island's airport and the tip of Garth's Nest. U.K. airplanes took off from the runways and began spraying toxic chemicals as soon as their wheels were up—in other words, at an altitude of thirty feet over adjacent communities.

One islander described her experiences with the spraying to me during an interview. Susan stood up from the couch and walked to the picture window that provided a panoramic view of the airport runways, which made it seem that they were in her front yard. As she stood there, the backlight from the window silhouetted her pregnant figure. "My husband and I stood here with our mouths open as we watched the planes release the dispersants as soon as their wheels were up, as they swooped over the house spraying the chemicals all over the window and roof. I can't believe they didn't wait until they flew over the wreck in Garth's Ness. There was a strong chemical odor in the house for days" (Norton, February 13, 1993). The 3,000 residents of Dunross Parish (located on the southern end of the islands) were not warned to stay indoors and were not informed of the times of the spraying. Nor were they evacuated from the region.

Dr. Derek Cox, a local public health official, told me that the 600 residents of the four communities were not evacuated because his office did not want to create an air of uncertainty or arouse public fear. However, Dr. Martin Hall, the Director of Environmental Health, told me that Cox initially was for evacuation until he spoke with Dr. Gerald Forbes of the Scottish Home Office. Public officials have often cited fear of public panic for their reluctance to release all the information about disasters. However, research by Mileti and associates has demonstrated that in the face of a disaster, most individuals do not become hysterical or irrational (1975).

By the time the spraying was terminated, almost all inhabitants on the western side of the southern end of the island had been exposed to both the dispersants and the oil, including eighteen pregnant women, most of whom were in their first trimester of pregnancy. The public health advisory issued by the Shetland Islands Council and by the local public health office simply advised residents to minimize skin contact with the oil and the dispersants and stay indoors. This advice was difficult to follow for the crofters who had to attend to their animals day and night during the storm and for the local school children who had to wait, twice daily, for school buses. Many of the other locals commuted to jobs, and hundreds of airport employees were constantly exposed to the dispersants while at work. Cleanup workers, who were issued rubber boots and slickers and masks, were concerned about their health. As with workers in the *Exxon-Valdez* oil spill, they often felt nauseated and ill. Only a small number of residents were provided with proper respiratory masks. For most, staying indoors did little good. Even indoors, they were nauseated by the odors that lingered in their homes for days after the immediate crisis passed.

Dr. Gerald Forbes, the director of the Environmental Health Unit at the Scottish Office, made a public statement that there was no health risk from the odors. Forbes contended that the tests conducted inside the homes revealed no measurable hydrocarbons. The homes, however, were not tested for the presence of other harmful chemicals that were in the oil and dispersants. Moreover, during the first ten days of the disaster, the health of residents was not clinically monitored. The lack of necessary equipment for monitoring and measuring pollution levels impaired the assessment of public health. Most of the monitoring that was conducted was for gaseous hydrocarbon levels. In the first few days, no blood or urine samples were taken of the exposed population, which was a critical oversight. Furthermore, oil droplets were not monitored, and equipment for the monitoring of chemical dispersants in the atmosphere was not available. The chemical content of the dispersants was unknown to both the people using them and those accidentally exposed to them. For the first forty-eight hours, not even the Island's health official knew the chemical composition of the dispersants. The content of the dispersants is generally a mixture of solvents and detergents, many of which are toxic. However, the level of toxicity of these compounds varies, and in some cases, the toxicity of a dispersant is largely unknown.

The Public Health Response

Once Dr. Cox was informed of the chemicals used in the spray, he withheld the information from all other officials and the general public because of what he referred to as *"commercial* confidentiality." As Dr Cox would later bemoan, his refusal to release this information and the chemical companies' insistence on withholding the information from the public were the largest factors for the pervasive suspicion of a conspiracy among the Shetlanders.

Eventually, the high winds and rough seas split the tanker asunder, and the oil appeared to wash out to sea, posing seemingly little threat to the islands. However, by the time the crisis had passed, all of the agricultural lands and the adjacent salmon farms were condemned because of contamination from both the oil and the dispersants. A 400-mile exclusion zone was established southwest of the island, banning the fishing and harvesting of salmon. This decision had a severe economic impact on ten boats and eleven salmon farms. A 1999 study reported that the exclusion zone was eventually lifted, but was still in effect for some species of shellfish (Goodland, 1999). These actions had significant economic impact not only on the seafood sectors, but also on the local fish packing and salmon processors (Moncrieff, 1993). Several million farmed salmon were destroyed. Salmon farmers and processors had issued a joint statement early on that they were strongly opposed to using dispersants in close proximity to any fish farms, fish spawning areas, and shellfish grounds. Although only the fish at one farm were deemed tainted, several major supermarket chains in the United Kingdom stopped purchasing the salmon.

> They have condemned some of the salmon and our season is ruined. My mates and I have lost our income for the season and we are worried that with the news of this spill Shetland Island salmon will be considered tainted for years to come and the global market will boycott our takes even if they are deemed safe for human consumption. What are we going to do then? Similar things have happened to other communities. How do we know the oil and dispersants won't harm the fish in the long run? Who knows where the oil will go. Look what happened in Alaska. (Scott, February 14, 1993)

While the international media corps went home, disappointed that the story was not as sensational as the *Exxon-Valdez*, an extended controversy began on the island that was nearly ignored by everyone but the local press. With complaints of too few dead animals to photograph and because of the severe weather that kept most reporters indoors, the world press ignored the human story that unfolded (Button, 1994a).

The Public Health Crisis

Almost immediately after the spraying of the dispersants, hundreds of Dunross Parish residents complained of eye, skin, and throat irritations; headaches, diarrhea, wheezing, coughing, chest aches, and fatigue. Over forty people experienced severe asthmatic responses. Shelly, a thirty-eight-year-old woman who lived on the south side of the island, had to wait outside for the bus up island to Lerwick were she worked in a small store:

> After several days I began to think I had the flu. It started with head and joint aches and eventually I had diarrhea so bad I missed a couple of days of work. I couldn't help but wonder if it was because of all the oil and chemicals. Standing outside waiting for the bus with the wind blowing so hard my face is covered with droplets of sticky, smelly chemical. I have to wash my hair constantly. Nothing like this has ever happened before. I think we all have a right to know what these chemicals are and how they might affect our health. Living

77

with the uncertainty of all this just is not fair. I think they need to be honest with us. (White, February 14, 1993)

Stephen is a burly, middle-aged man who has tried making a living doing many different things on the island. He gave up being a mechanic so he could continue the traditional life as a crofter. His response was similar to many other people on the island exposed to the effects of the spill.

> I have to go out during the day and attend to our sick livestock, there is no way I can avoid it. They say we should stay inside as much as possible, but how are we to make a living if we do that? Anyway, the smell of the oil and chemicals is almost as nauseating in the cottage as it is outside. Ever since the first couple of days my eyes are red and swollen and so are my wife's and she has tried to stay inside. I have constant headaches and a cough that plagues me day and night. (Whittaker, February 15, 1993)

In the weeks ahead, more than 250 people would demonstrate abnormal lung functioning. A smaller number would have test results that demonstrated renal and liver malfunction. Many of the local residents became alarmed. When Greenpeace and an environmental group from Norway disseminated literature that criticized the use of dispersants, residents grew more uneasy. Greenpeace also pointed out that dispersants had never been approved for use around humans and had unpredictable biological effects. The environmentalists cited reports that dispersants may adversely affect the amount of oil absorbed by the lungs and increase the rate of absorption of toxic oil by the mucous membranes of the mouth and nose. One of the dispersants used on the spill (Dispolene 34s) was not licensed for use in the United Kingdom. Another dispersant was criticized because it had never been approved for use along rocky shorelines. Yet another of the chemicals was banned in Norway because it failed toxicity tests. U. K. agriculture minister David Curry did not admit until January 21 that one of the dispersants had failed tests for use on rocky shorelines and that another dispersant had not even been tested for use along rocky coasts because it had not been developed for such usage. Finally, he stated, "The disclosure of this information underlines that residents of the south Mainland of Shetland were well justified in expressing misgivings about the use of dispersants."

Disputed Information

The man who immediately found himself immersed in the middle of this controversy was Dr. Chris Rowlands, the general practitioner for the Parish. In contrast to the assurances that all was well given by Cox, Rowlands advised school officials not to allow school children outside during recess. He also informed a public gathering that the crude oil contained three chemicals known to cause cancer: butadiene, naphthalene, and benzene. He questioned the use of dispersants, telling another public group that U.K. officials were keeping the chemical contents of the dispersants a secret even from him, the parishioners' physician. Moreover, he alleged something that later was revealed to be true—that the United Kingdom was using an old, outdated stock of dispersants, and that the government was unsure of their contents because barrels of different chemicals were mixed together.

Indeed, the dispersants were so old that a number of the barrels stored at the Sumburgh Airport leaked and created their own spill problem. One of the guards who worked at the airport informed me that some barrels were old and rusty. A press release issued by the Shetland Islands Council stated that "it is not known how much has leaked but it could potentially be 500 gallons" (January 27, 1993). The spill caused an uproar on the island and further damaged the reputation of the MPCU.

According to a number of informants, including officials of the Shetland Islands Council, there was a considerable concern about secrecy and conspiracies. For example, under Scottish law, meetings are to be conducted in public; however, there is a process whereby public meetings, including those of the council, can be closed if commercially sensitive matters are being discussed. Under this veil of secrecy, the contract with BP to construct and operate the Sullom Voe facility was negotiated. Even now, the general public does not know the conditions of the contract. According to Willis (1991), "[V]ery large subsidies from council funds are handed out in secret." These conditions contributed to the public fears during the spill crisis.

A British Petroleum employee told me that BP sent their employees (BP operates the Sullom Voe oil terminal and employs many people) a memo that told them that BP was going to "keep their heads down" and make no comment on the spill. Because so many BP employees live in the Shetlands, the rumor of this memo spread rapidly. A number of people told me that they interpreted this as yet another indication that a conspiracy was astir. Although it had nothing whatsoever to do with the oil spill, BP was the manufacturer of one of the dispersants.

Alarmed over the symptoms many of his patients were exhibiting, Dr. Rowlands decided to begin clinical monitoring of his patients two days after the spill, at a time when the United Kingdom and the local health board refused to conduct medical monitoring of the residents. Dr. Rowlands began taking blood and urine samples of the parishioners, especially those whose occupation forced them to spend most of their time outdoors. In an effort to conduct thorough tests, he phoned Campbell of the Scottish Health Board and asked him to begin tests on all the residents in the exposed regions. According to Dr. Rowlands, as eager as Dr. Campbell was to conduct the tests, he ran into severe opposition in the Scottish Home Office. However, once the press revealed that Dr. Rowlands was single-handedly conducting tests, the Home Office relented. Unfortunately, because of the time delay—at least ten days after the spill—the tests were almost meaningless; moreover, tests for carcinogens such as benzene, toluene, and zylene were never conducted.

Both Home Office officials and officials of the Shetland Islands Council pressured Dr. Rowlands to stop making public statements and conducting tests. Dr. Cox made public statements to the press that Dr. Rowlands was using scaremongering tactics and creating public hysteria. Cox also accused both Greenpeace and Bellona, a Norwegian environmental group, of being more of a public health hazard than the spill by causing unnecessary alarm. In an interview with me, Dr. Cox blamed the unrest on inflammatory statements made by "some individuals or groups from Alaska that told our people" that the public officials in Alaska withheld information and that U.K. officials would take the same tactic and must be challenged (Cox, February 8, 1993).

Community Response

Amid this controversy, the islanders themselves were very concerned about the lack of testing and the use of dispersants. A petition, signed by several hundred people, to the Shetland Islands Council demanded, among other things, that there be an open investigation of the disaster and that all information regarding the use of chemical dispersants be disclosed to the public. Salmon fishermen and fish processors took a stronger position and issued a joint statement opposing the indiscriminate use of dispersants.

According to Martin Hall, director of Environmental Health, even members of the Shetland Islands Council were upset over the use of dispersants. However, when they protested to the MPCU, they were told, "I don't care if you like it or not, we are going to do it" (Hall, February 9, 1993). The controversy over the spraying of dispersants came to a head when residents in Scatnes—a community immediately adjacent to the airport—and several airport employees threatened to sit down on the runways and block planes from taking off. Malcolm Green told me that, at this point, he informed the U.K. Department of Transport that he wanted the spraying discontinued. As a result of this demand, a compromise was reached whereby spraying would be limited to certain areas and allowed only at certain altitudes (Shetland Islands Council Public Notice, January 10, 1993).

The official word from the Scottish Home Office from the first day of the spill was that there would be no long-term health effects from either the spill or the use of dispersants. This claim was largely unsubstantiated. This spill was unique in that it was the first-known instance of respirable oil droplets, rather than vapor, being blown ashore; there were no preexisting studies of the possible health hazards of such an event. To make matters worse, chemical dispersants were being used in unauthorized areas and contaminated nearby settlements.

Furthermore, although some studies suggest dispersants increase the danger of oil inhalation, we have no scientific knowledge of what effects the combination of oil and dispersants have on the human body. The only certainties are that both the oil and chemicals contained known carcinogens. Under these circumstances, accepted scientific practice would require the assumption of a risk, preventative action, and monitoring of the environment and the human population.

Not until eleven days after the wreck did the Scottish Office begin a limited medical survey of the population. Out of 640 individuals invited to participate, 460 were tested, along with a control group of ninety-six individuals from the northern end of the island. The report showed no significant differences between the two populations. However, the tests were conducted far too late to detect toxic chemicals present in the body during the spill and far too early to detect cellular abnormalities. Furthermore, no tests were conducted to monitor known carcinogens such as benzene (Fogg, 1993). In keeping with the secrecy surrounding the spill, Dr. Rowlands was not shown the test results until two hours before the press conference at which the results were announced. Given such little time to review the report, Dr. Rowlands was effectively prevented from making a meaningful contribution to the press conference and to the public's awareness.

Drs. Campbell, Cox, and others eventually began a study, which was, in some ways, too little too late. Toxicological sampling did not begin until eight days after the tanker went aground. The investigation did offer insight into some of the short-term health effects (Campbell, et al., 1993). However, as the study admits: "The lack of sufficiently detailed air monitoring until day 5 precluded the examination of any relation between symptoms and possible exposure dosage. The post-code information collected lacked sufficient precision for attempting to determine a relation between symptoms and distance from the Braer." (Campbell, et al., 1993).

Another shortcoming of the study is that toxicological sampling did not begin until eight days after the tanker went aground. The study revealed that the exposed population presented symptoms traditionally associated with hydrocarbon exposure. Nonetheless, it found that no laboratory evidence of acute health effects could be determined within the parameters of the study. Ironically, the report concludes that "Rapid epidemiological responses to environmental incidents are necessary and feasible." (Campbell, et al., 1993)

The same authors later issued a report from a study that was conducted in June 1993. The participants in the previous study were given an extended questionnaire. A systematic sample of exposed people who had not participated in the previous study was surveyed over the phone.

> Comparison of the symptoms of exposed people in the two weeks before this phase with their presence immediately after the incident showed more tiredness, fewer throat, skin, and eye irritations and fewer headaches in June than after the event. In June exposed people were more likely to report throat irritation and breathlessness on exertion in the previous 14 days. A greater proportion of those exposed reported weakness at some time over the five months. Eighty-six per cent declared these symptoms to be persisting, with one third describing their onset in January. (Campbell, et al., 1994)

In some ways, official monitoring of animals was better than the monitoring of people. Sheep and cattle that had died during the disaster were autopsied, and biopsies were performed on the tissues of animals that were ill. As with the human medical situation, however, the test results were not released to the general public or even to the veterinarian of Dunross Parish. The veterinarian was as unsuccessful as Dr. Rowlands at obtaining information for his clients.

Cultural Considerations

The human world, as Rappaport (1988) reminds us, does not consist merely of chemical and biological processes alone, but is saturated with meaning and value. Cultural conditions predate the North Sea oil era. For example, attitudes about off-islanders figure predominately in the suspicion and distrust many islanders exhibited during the crisis. Native Shetlanders have a deep attachment to their non-British history and many identify strongly with their Scandinavian ancestry. Both the English and the Scottish nationals who arrived en masse after the discovery of oil to work in the expanded economy were and are viewed with suspicion as *off-islanders*, people who are insensitive to the Shetland way of life and who represent British neocolonialism.

More important, the crofts and the crofters are viewed by island natives as symbolizing an ancient way of life. Crofts are viewed as land to which people have an inalienable right. They were originally protected in the larger context of Scottish Nationalism in the 1886 Act that made landlords powerless in making decisions that affect the use of crofting lands. While this special status was revoked in 1911, a new Crofters Act was produced in 1955 after a government report (The Taylor Report, 1954) argued for the protection of crofting communities because "they embody a free and independent way of life which in a civilization pre-dominately urban and industrial in character is worth preserving for its own intrinsic quality" (Parman, 1990). While today crofting is not a viable economic project for most people, it is, nevertheless, an important way of life (Cohen, 1987).

The crofts are also landmarks that resonate with great symbolic significance because some families have resided in a particular croft for generations. The cottages of crofters are commonly built on or around the remains of earlier dwellings that predate them by as much as several thousand years. Crofts are, in a sense, territory for a lineage. The croft also serves as an aggregate social identity and has traditionally been the core of fishing crews. Individual identity is often permanently linked with the croft, even if a crofter lives in other parts of the archipelago. Most topographical features of the crofts have personal names associated with their previous inhabitants. Thus, crofts have become "historical repositories" of family lineages and the islands' cultural history (Cohen, 1987).

Seen from this perspective, the spraying of the crofts and their moors and the contamination of these croft dwellings, lands, and topographical features were viewed by many as a violation perpetuated by an off-island, industrialized society. The inalienable right of crofters to the land was violated by the spraying of crofts and the refusal of U.K. officials to inform crofters of the contents of the chemical solution that befouled both their economic livelihood and their cottages. Thus, the contamination and condemnation of the crofters' animals and lands threatened their way of life and demonstratively desecrated the cultural heritage of the islands.

This discussion illustrates that, when public authorities assess risks, they need not only to consider the physical and psychological risks that accompany an oil spill or the use of dispersants, but the risks to the social and cultural fabric of a community. Recognizing that trauma inflicted by disasters has an impact on our socioecological system as a whole is imperative.

Discussion

Illustrative of the way many technological disasters are handled, the response to the wreck of the *Braer* was to use a quick technological fix (Button, 1994b). The solution to the problem was seen as an engineering issue; thus, the human dimensions of the disaster were relegated to the background and all but ignored. The human component was eventually foregrounded when local people protested the indiscriminate use of dispersants and the disregard for their health and safety.

Why did Dr. Cox, the director of Public Health, risk the credibility of his office so early in the disaster and refuse to release information that would enable victims to make their own risk analysis? The answer lies in at least two realms. In order to come to some *rational* judgment about the efficacy of using dispersants, Dr. Cox felt that he needed to know the chemical composition of the sprays. However, the manufacturers of the dispersants would not release the information unless it was under the cloak of *commercial* confidence. In talking with Dr. Cox I clearly learned that he personally viewed retention of this confidence a matter of great professional integrity. Constrained as he was by his own code of honor, he was also restricted by laws and policies of the U.K., which, despite being one of the seats of the democratic tradition, has no right-to-know policy as formulated in the United States.

Access to Information

The European Community (EC) has established procedures for the communication of hazardous information, which are distinctly different from the approach in the United States, which is predicated on a federally mandated right-to-know law. The European Community has an information policy named after the tragic Seveso chemical accident in Italy in which several communities north of Milan, including the residents of the town of Seveso, were exposed to dioxin. The Seveso Directive (1979) established a framework that delineates responsibility of both industry and government to communicate hazardous information to the public. Industry is required to provide detailed accounts of toxic materials to a public authority; however, unlike the policy of the United States government, the EC government authorities are not required to release the details of this information to the public (Collins, 1992).

While both the United States and European Community countries regard information as an important resource, European legislation and policies interpret public access to hazardous information based on a need-to-know rather than a right-to-know policy (Eijndhoven, 1994). Instead of a policy of an open-ended access to all information, the Seveso Directive provides access only for specific information that is needed for a specific purpose. Within the European Community, countries have varying policies regulating the amount of access their citizens can have to risk data. The United Kingdom has the most restrictive approach in the Western world in withholding information from the public. U. K. policies emphasize providing scientific advisors with risk data and allowing them to make recommendations in the interest of the public, just as they did in the case of the Shetland Islands spill. Off-the-record comments given to me by both Shetland and U.K. officials strongly suggest that Cox was under considerable pressure to minimize public concern and to protect corporate interests.

By March of 1993, Cox adopted a different posture. In a conference report he stated:

> In this incident, the manufacturers of the chemical dispersants used would release the complete chemical composition of their dispersants

to public health authorities only on the understanding that it would be held in commercial confidence. I cannot think such a confidence was necessary. The chemicals used were not themselves secret in any way, and it is difficult to believe that commercial rivals are not themselves capable of analyzing their competitor's products. If it had been possible to make the composition of these public it would have been relatively easy to reassure the public about the safety of their use, but because authorities dealing with the incident were unable to release this information not only was there an unjustifiable public concern about the use of these dispersants, but the secrecy surrounding their use and composition seriously undermined the confidence of the public in the pronouncements made about the relative lack of risk to the public from the incident as a whole. The withholding of this information was the most serious hindrance to handling the public health dimensions of this incident, and did no credit to the companies concerned. (Cox, March 31, 1991)

Medical Uncertainties

Contrary to Cox's assertion, there still would have been controversy even if the information had been released to the public; the decision to use dispersants was undoubtedly unwise and unnecessary. Furthermore, the lack of adequate monitoring of both the environment and humans was unthinkable given the unique conditions surrounding the spill. No doubt, however, the air of controversy would have been greatly diminished if vital information had not been withheld from public scrutiny. The public had a right to ascertain for themselves the risks involved in the affair—their access to the information was critical. The refusal to release the information seriously eroded the credibility of the public health office to the point that victims became mistrustful of and hostile toward the office and increasingly suspicious of the circumstances surrounding the spill. The public clearly perceived that the interests of corporations were, in this instance, placed above the interests of the public.

Considerable uncertainty was generated after the Shetland Islands oil spill, just as it was after the *Exxon-Valdez* oil spill (Button, 1993) and the Santa Barbara spill (1969) (Molotch, 1970; Easton, 1972). As in these other spills, the climate created by the withholding of information generated increased uncertainty about the credibility and trustworthiness of officials: uncertainty is often "equated with surreptitiousness and incompetence" (Marrett, 1981). Moreover, increased uncertainty only adds to the number of interpretations that can be made about a disaster. In this case, as in most, the knowledge that was withheld was not only a key to meaning, but in very real terms a key to power (Keesing, 1987). People construct their social realities by making interpretations and conferring meaning based on knowledge. Access to the facts is crucial. People employ knowledge in order to cope; they use knowledge not only to interpret events but to act. The limited access of the Shetland Islanders to knowledge left them in a precarious position and decreased their ability both to cope and to act. Both the U.K. disaster response team and the Public Health Office failed to incorporate into their response the perceptions and reactions of the local people. By not providing community members with hazard information, they denied the victims the knowledge with which to interpret and respond to the event in a meaningful manner. By withholding information, officials narrowed the number of informed participants who could participate in the debate.

If a lesson is to be learned in this case, it is this: The local perception of risk is mediated by the perception of how risk is being managed by officials. As important as the disaster event itself is the way it is perceived and interpreted by the community. Anxiety is minimized if the risk is perceived as being adequately managed and remedied. However, if local residents become distrustful of local government, local concern over risk will increase (see Fitchen, et al. 1987; Edelstein, 1988; Button, 1993). The risk management strategies of the Unted Kingdom and the local public health office failed to take this into consideration and, thus, not only added to the air of uncertainty and anxiety, but made a crucial difference in whose interpretations were heard and whose interpretations were excluded.

The Sea Empress Spill

Unfortunately, this lesson was not learned. Two years later, in the wake of the *Sea Empress* spill in Wales, similar health concerns and uncertainty arose.

On the night of February 15, 1996, the Liberian-flagged, Russian-crewed 147,000-ton oil tanker *Sea Empress* went aground on the rocks in the mouth of Milford Haven, Wales. By the time the tanker was freed six days later and towed into harbor, the ship had spilled over 90,000 tons of crude oil onto the Pembrokeshire coastline, one of Britain's most pristine coastlines and its only national park. The spill was approximately twice the size of the *Exxon-Valdez* oil spill and the second largest to occur in

Great Britain's waters. Ironically, although the spill was larger than the *Braer* spill, it received much less worldwide publicity. Although the *Sea Empress* disaster approached the size (100,000 tons) of the 1967 spill of the *Torrey Canyon*, which was also headed for Milford Haven and was at the time described by Great Britain as its largest peacetime threat, the Thatcher government refused to recognize the *Sea Empress* as a "disaster."

As in other oil spills, spill response authorities paid little attention to the anxiety and uncertainty about public health issues. In the town of Dyfed, the director of public health had to rely on radio and television accounts for information and was not even officially notified of the spill until the day after the accident. Five days into the spill, he had to break into a meeting of the *Sea Empress* response team in order to obtain information. Formal ongoing monitoring of the atmosphere and the ocean did not commence until the third week in March. Local public health directors advised strict supervision of children on the beach and warned people to avoid contact with the oil. People were also warned not to collect or consume local shellfish. In the Milford Haven area, residents were warned to stay inside, close their doors and windows tightly, and avoid inhalation of the oil fumes—an impossible task, even indoors. A number of local people complained of respiratory problems and other ill effects associated with exposure to hydrocarbons.

Despite efforts of some local health officials, the actual monitoring of the event was insufficient and too late to be of any significance. Four weeks after the spill, the Dyfed Public Health Office mailed two thousand questionnaires to area residents about their perception of possible short-term and long-term health effects of the disaster. The findings of the retrospective cohort study were that there was an increase in the reported prevalence of headaches and sore eyes and throats after the first day of the accident, "suggesting a significant direct health effect on the exposed population" (Lyons, 1999). The exposed population reported an increase in self-reported physical and psychological symptoms in the four weeks following the incident. The exposed population also reported significantly more anxiety and depression. The study's authors concluded that the exposed population's physical health remained significantly worse than the control population. Again, the Dyfed Public Health Office study offered the caveat that "had data on individual exposures been available, a dose response relation would have strengthened this conclusion" (Lyons, 1999).

4. "Damaged by Katrina, Ruined By Murphy Oil"

On the morning of August 29, 2005, Joe Navis, thinking the eye of Hurricane Katrina was about to pass over his home in St. Bernard Parish, tried to stay awake as he listened to the radio for news of the approaching storm. Eventually, Joe drifted off to sleep. Twenty minutes later, the sound of an explosion awakened him. He leaped from his couch. Water pouring into the room through the French doors was as high as the doorknobs. Instantaneously, he heard the windows shatter as the water started pouring in from all directions:

> I said, "Dad, we have to get out of here, we're going to die." The water was rising higher and higher. By the time we got out of his room, we had to swim under water. So, we just went all the way under to get to the door frame. … Dad swims around the corner to the front door and he realizes he can't see light coming through the door. This all happened in a matter of minutes and then all of a sudden he says, "Son, we gotta get out of here, we're going to die." He tried to push the door, but he couldn't get it open because the water on the outside was higher than on the inside, and the pressure of the water was amazing. "We're trapped!" Dad yelled. Then the door blew open and we were literally swept away into a river of water. (Navis, February 26, 2006; see also Young 2005)

The sea storm surge entered Lake Borgne. From there, it sped down the path of least resistance—the Mississippi River Gulf Outlet (MRGO), which came to be known as "the highway of death." The surge then smashed into the industrial canal, creating an enormous breach and

"Damaged by Katrina, Ruined by Murphy Oil."
Photograph © Gregory Button.

inundating the Lower Ninth Ward and St. Bernard Parish's seat, the city of Chalmette. The area of the Ninth Ward near the breach was completely devastated; six feet of water flooded the neighborhood. In St. Bernard Parish, the water rose nearly twice as fast, eventually reaching heights of ten and even twelve feet. It demolished many structures in its path, pushing huge fishing boats inland and flinging cars onto the tops of homes. An area of 680 square miles of St. Bernard Parish was flooded, all but obliterating the community. Among the tens of thousands of homes and commercial buildings, only two were left untouched. Over 2,700 hundred homes were left entirely underwater. Sadly, with 129 people dead, the Parish also suffered the highest mortality rate of any Gulf Coast community.

As if this tragedy were not enough, the worst was yet to come. A disaster within a disaster was inflicted on the community when the storm waters shoved a 65,000-barrel Murphy Oil refinery tank off its moorings and moved it more than thirty feet, spilling 1.1 million gallons of toxic crude into the adjacent communities of Meraux and Chalmette. In the event of an approaching storm, oil refineries are supposed to top off their tanks to lessen the likelihood that a storm surge could topple them. Murphy Oil neglected to do so. The thick, murky oil swept into homes. Once it receded, it left oil lines on the houses as high as six feet, along with a slurry of toxic chemicals. Joe Navis's family home was among the 1,800 affected by the spill.

The oil spill materialized one of the most persistent fears of the residents of the Parish. One resident told me, "My family and I have lived our whole life in the shadow of the refineries with the constant fear that either the plant would explode or that there would be an oil spill that would flood our neighborhood" (Mackin, February 27, 2006).

A History of Disaster in St. Bernard Parish

Although Hurricane Katrina was the first occasion during which St. Bernard Parish experienced this kind of double calamity, it was, by no means, the only disaster to affect the community in recent history. In 1965, Hurricane Betsy struck the Parish hard and damaged nearly half of its homes. Then, in 1969, Hurricane Camille heavily damaged the lower half of St. Bernard Parish (Din, 1998). A refinery fire in January 2001 was followed just a couple of years later by another accident in which a train hit a tanker truck as it was pulling out of a refinery, igniting a fire and killing three people. A study released in 2006 revealed that the

Exxon/Mobil Chalmette refinery had seventy-eight accidents in a thir-teen-month period.

The disaster that is perhaps most salient to our interpretation of the events surrounding Hurricane Katrina is also one that many people see as emblematic of the marginalized status of parish residents. During the great 1927 Mississippi flood, powerful development interests in New Orleans prevailed in gaining permission to dynamite a levee at Poydras Plantation. The act was supposed to alleviate the potential flooding of New Orleans, although most doubted that it served any purpose at all. The residents of St. Bernard Parish were mostly hunters, trappers, and fishers, most descendents of Islenos, people who came from the Canary Islands in the 1700s and were among the first Europeans to settle in Louisiana. Joe Navis's ancestors were among these settlers. The occupants of this swamp land were coerced into allowing their land to be inundated and were promised financial reparation in turn. The blasting of the levee dev-astated the Parish, destroying livelihoods and homes, but the residents were never compensated for their losses and the economy of the Parish suffered for years (Daniel, 1977; Barry, 1997; Kelman, 2003). As the levee was dynamited, St. Bernard Sheriff L. A. Meraux echoed the feelings of many: "Gentlemen, you have seen today the public execution of this par-ish" (Barry, 1999: 257). This event has endured in the memory of Parish residents to the present. As a result, many parishioners perceive their mar-ginalized status in the context of industrial pollution. At the same time, citizens retain a lot of pride in their working-class enclave. Spray painted on the roof of one building in the downtown area is "Chalmette, Salt of the Earth."

The Highway of Death

Another economic and environmental blow to the local economy and culture that influenced perceptions of Katrina was the construction of the MRGO. This development project created what has become known as "the highway of death" because it ultimately made the area vulnerable to the wrath of Hurricane Katrina. Initiated in 1958 and completed in 1965, MRGO is a seventy-five-mile-long industrial canal that provides a shortcut shipping channel from the Gulf to downtown New Orleans. The banks of the waterway are constantly eroding and widening the channel. In some places, the channel has increased from 650 to almost 2,000 feet in width. Meanwhile, the waterway lost 87 percent of its traffic during the last three decades; it now serves less than three ships per day at an expense of $16 million per year to dredge and maintain (Cooper and Block, 2006).

The canal has destroyed thousands of acres of marshland as well as an ancient cypress stand and has promoted salt-water penetration of the marsh, destroying an estimated forty square acres of marshland every year. The salt-water intrusion ruined the muskrat trapping, long a mainstay of locals, and lowered both fish and oyster yields (Din, 1988). According to Sid Navis, the canal "ran all the muskrats out and destroyed the fur trade," on which he and many other locals depended for their economic liveli-hoods (February 26, 2006). As Joe Navis proclaimed, "MRGO killed it [the fur trade], it put the final death knell on it" (February 26, 2006).

The New Orleans District of the U.S. Army Corps of Engineers speculates that the loss of land in the area approaches nearly 3,400 acres of fresh/intermediate marsh. More than 10,300 acres of brackish marsh, 4,200 acres of saline marsh, and 1,500 acres of cypress swamps and levee forests have been destroyed or severely altered (Proctor, May 2009). The U.S. Department of Interior issued the following ironic statement in 1958: "Excavation of the (MRGO) could result in major ecological change with widespread and severe ecological consequenc-es" (Freudenburg, et al., 2009: 86).

After Hurricane Katrina, the demands of many local elected offi-cials and citizens for the closing of MRGO escalated. "MRGO has got to go" became a widespread chant. Finally, in July 2007, the Army Corps of Engineers recommended to Congress the total closure of the structure (Rioux, July 3, 2007).

Cancer Alley Doesn't Stop at New Orleans

Traditionally, "Cancer Alley" refers to an eighty-five mile stretch of pet-rochemical plants and refining industries that line the Mississippi River from Baton Rouge to New Orleans. For many citizens, this definition falls at least five miles short. Ken Ford, a lifelong resident of the city of Chalmette and founder of St. Bernard's Citizens for Environmental Quality, has often proclaimed, "Cancer Alley doesn't stop at New Orleans" (Ford, 2006). Since World War II, the city of Chalmette in St. Bernard Parish has been home to four large oil refineries whose massive, sprawling structures bump up against the fence lines of residents and fill the night sky with an eerie luminescence. Ken has lived adjacent to the Exxon-Mobil refinery for forty-five years. He has endured both lung and bladder cancer, and he is not alone in his suffering. According to the Louisiana Tumor Registry, St. Bernard Parish has the highest cancer mortality rate of any parish in the state. For white females and black males, lung cancer rates are 35 percent higher than the state average; the rate for white males is 22 percent higher (Miner, 2004: 1). On one block of Ken's street alone, fifteen people have died of cancer. The battle that Ken and his neighbors are fighting is more elusive and prolonged than the Battle of New Orleans of 1814–1815, which was fought less than a mile from his front door. Ironically, the British troops who died in that battle are literally buried beneath the Exxon-Mobil refinery.

Industrialization and Activism

The first oil wells in Louisiana were drilled near Baton Rouge at the turn of the century, and commercial development increased in subsequent decades. By the 1920s, oil wells were being drilled throughout most of the state. The Mississippi River chemical corridor quickly became one of the largest chemical and refining centers in the nation (Markowitz & Rosner, 2002). The history of oil and the petrochemical development in Louisiana is markedly different from that of other states. Multinational oil companies formed an early and effective alliance with the powerful and wealthy planter-merchant alliance that successfully disenfranchised blacks and poor whites in the state constitution of 1898.

This alliance between powerful multinationals such as Standard Oil and the state government, the political economy of the state, and a lack of government transparency fostered the massive development of the industry and its disregard for the region's environment and its people's health. However, it also gave rise to a formative environmental justice movement (Allen, 2003). Grassroots activism became one of the few ways in which local residents could challenge the powerful industrial/governmental alliance of oil.

Beginning in the 1940s, urbanization and industrialization radically transformed St. Bernard Parish and the city of Chalmette from an economy largely dependent on hunting, trapping, shrimping, and agriculture. Kaiser Aluminum built a plant in Chalmette. Tenneco Oil established the refinery that became the Exxon/Mobil refinery. Chalmette and Murphy Oil companies also emerged on the industrial landscape. Along with these developments came a large, white, working-class population fleeing downtown New Orleans and seeking the newly created jobs and postwar housing (Campenella, 2006). In the span of thirty years, the city grew from a population of some 7,000 to over 67,000 residents. Then, by 1980, this period of economic growth collapsed due to a stagnating regional economy. The Kaiser Aluminum plant closed, and the largely working-class community endured hard times that lasted until Hurricane Katrina made landfall. Then residents faced even harsher times.

The industrialization of the community and the resulting inequitable distribution of power and wealth led to conditions similar to those in Cancer Alley. The rampant development of the refinery industry brought massive environmental pollution and sharp increases in the incidences of cancer and respiratory and cardiac disease. The only major difference is that the affected community was largely white rather than African American. Nevertheless, a strong grassroots activist and environmental justice movement emerged just as it did up and down the chemical corridor from Baton Rouge to New Orleans. Within the last decade and a half, groups such as the St. Bernard Citizens for Environmental Quality, The Louisiana Bucket Brigade, Fence Lines, the Sierra Club, and Concerned Citizens around Murphy have become active in the community.

Prior to the Murphy Oil spill, local groups were concerned with the industrial pollution of the refineries that abut their schools and backyards. Residents were, and remain, concerned with air and water pollution and industrial accidents, of which there have been many in recent years. Periodically, flame plumes light up the night sky, signaling emergencies; however, because of their regularity, these plumes have become part of the skyline. According to the Tulane Environmental Law Clinic, refinery company reports state that substances considered toxic by the U.S. Environmental Protection Agency (EPA) are routinely discharged—about one million pounds of pollutants in excess of permitted limits per year. Residents complain of foul smelling chemicals that pollute their homes, schools, and workplaces. Ken Ford, community activist, says that Exxon has been monitoring the site for years but with "ineffective monitors placed two miles away" (Ford, 2006). Exxon/Mobil has been using monitors that measure in parts per million even though in Louisiana the ambient air standards are suppose to be measured in parts per billion. The Louisiana Department of Environmental Quality (LDEQ) installed improved monitors closer to the refinery areas, but Ford believes the agency needs to install additional monitors downwind. The Louisiana Bucket Brigade and the St. Bernard Citizens for Environmental Quality have been working together to study the potential of these toxins for long-term environmental health problems.

The Bucket Brigade trains community members in the use of an EPA-approved device that consists of a five-gallon bucket and a hand pump vacuum on its lid. The data collected is sent to independent testing labs. According to Anne Rolfes, the head of the Bucket Brigade, the test results tell a story different from that of Exxon/Mobil's monitors. Rolfes claims that her group's tests show several sulfurs—many of which are known to cause respiratory diseases and are carcinogens. Rolfes says that these findings, along with the presence of rare cancers in the neighborhoods, is "at least anecdotally shocking" (Rolfes, et al., 2003).

In 2003, the two groups, with the aid of the Tulane Environmental Law Clinic, filed a suit for violations of the Clean Air Act and the Emergency Planning and Community Right to Know Act. Their complaint alleges that Exxon/ Mobil's Chalmette plant violated both acts. The suit specified the following violations:

1. Emitting quantities of sulfur dioxide, nitrogen oxide, nitrogen dioxide, volatile organic compounds, and other harmful pollutants that far exceeded the refinery's permitted levels;

2. emitting quantities of benzene (a cancer causing substance) that far exceeded the refinery's continuous emissions limits for that substance;

3. failing to properly maintain the flares that burn off dangerous emissions;

4. failing to file reports required by law each time an illegal emission occurs. (Rolfes, et al., 2003)

Anne Rolfes stated: "We are suing Exxon/Mobil because their violations of the law threaten human health. For the past several months Exxon/Mobil has refused to give us any information or even respond to our letters" (Rolfes, et al., 2003).

In 2005, a federal judge ruled that the refinery violated the Clean Air Act more than 2,660 times since 1999 by discharging more pollutants than their permit allowed. It was also found guilty of surpassing allowed emissions of benzene by more than 1,200 times, for exceeding the limits allowed for the release of sulfur dioxide (563 violations), and for the release of excessive flare ups (802 violations).

The Aftermath of Hurricane Katrina

After Hurricane Katrina, attention shifted to the Murphy Oil spill. In the immediate aftermath, the U. S. Coast Guard took charge of the initial clean up by overseeing Murphy Oil's removal of oil in canals, the tank containment area, and storm drains. The workers repaired the damaged container tank and attempted to skim oil from the surrounding water. The National Guard cordoned off the area and prevented residents from reentering the neighborhood. The responsibility for supervising the cleanup fell to the U. S. EPA, which worked in conjunction with the LDEQ. After the waters receded, some oil evaporated, but much of it remained on streets and in the soil and permeated the exteriors and interiors of homes. With time, the oil seeped farther into the ground, making retrieval all but impossible. The question became, who would be responsible for cleaning up the remaining oil from the ground and the existing infrastructure and how would they achieve this task? Murphy Oil informed the residents that the crude oil "will not present short-term or long-term health or safety concerns" (Louisiana Bucket Brigade).

Scientific Uncertainty in the Wake of Katrina

Following most technological disasters, definitive, conclusive, and scientific evidence about the potential health effects of toxic chemicals is lacking. Clearly calibrated safe exposure levels and specific correlations with specific disease entities are rarely available. In cases of specific correlations with disease, there are controversies about what constitutes safe levels of exposure as well as which diseases may surface over time. Usually, such events generate a climate of controversy within the scientific and lay communities. As Van Loon has astutely observed, the crisis of uncertainty often becomes a feedback loop of uncertainty: "More uncertainty demands more knowledge, more knowledge increases complexity, more complexity demands more abstraction, more abstraction increases uncertainty" (2002: 41).

In such circumstances, citizens quickly become disillusioned about the reliability of scientific evidence. Science is perceived as lacking certainty and incapable of predicting the future or offering assurances. While corporations and public agencies reassure the public of the certitude of scientific findings, health and safety issues are often contested. Frequently, disaster victims perceive uncertainty as indicative of either incompetence or surreptitiousness. In the past, officials and corporate spokespeople have argued that what is perceived by some as a *disaster* is a *non-event* or an event of no long-term consequences.

For example, following the accident at Three Mile Island, perceptions about what actually happened were diverse and conflicting. While some people called it a mishap, others were certain it was a catastrophe (Nelkin and Brown, 1982). There were similar disputes about many other environmental disasters including the events at Love Canal (Levine, 1982), the contaminated groundwater in Woburn, Massachusetts (DiPerna, 1985), the *Exxon-Valdez* oil spill (Button, 1993), and the Shetland Islands oil spill (Button, 1995). The Murphy Oil spill followed this pattern. The EPA and LDEQ tried to minimize the impact of the spill. They underplayed any potentially long-term impact and reassured the public with their scientific certainty: a certainty based on scant and dubious evidence.

The residents of St. Bernard Parish, like many victims of Hurricane Katrina, were and are forced to deal with multiple uncertainties, including the uncertainty of being able to repair or rebuild their homes. The most pervasive uncertainties for St. Bernard Parish concerned environmental contamination and the health and safety of their families and community. However, recognizing that risk assessments are not made in a vacuum is important. While there are multiple environmental risks and uncertainties, there are also other uncertainties such as housing, the safety of FEMA trailers, jobs, insurance settlements, access to health care and schools, and individually specific concerns. The risk assessments and coping strategies of residents of St. Bernard Parish must be seen in this larger sociopolitical perspective and not only, as scientists and academics often analyze and discuss, within the confines of traditional risk assessment discourse. The stochastic models of risk assessment cannot begin to address the bewildering array of uncertainties with which residents of St. Bernard Parish must contend. The narrowly defined approach of risk assessment models cannot adequately address the complex calculus of decisions that confront residents.

Residents Face Doubt and Uncertainty

As the community predicament became severe, community residents began expressing serious doubts about such reassurance. One resident asked, "What are the health risks—aside from the respiratory problems everybody seems to have? Will we get cancer five years from now from the Oil?" (St. Bernard Parish Resident, March 18, 2006).

Joy Lewis of the St. Bernard Citizens for Environmental Quality asked, "What is going to happen to us, are we going to get sick from returning and breathing all that [oil]?" (Lewis, March 17, 2006).

Another resident stated: "I am four months pregnant and I have two toddlers; how can I be sure it is even safe to return to my home and assess the damage, let alone return to live? Who can honestly tell me if it is safe or not?" (St. Bernard Parish Resident, March 18, 2006).

A homeowner declared: "Nobody says anything because they want you to come back. This is cancer alley all along the river—that's why people don't want to bring their kids back. I don't even want to grow vegetables here" (St. Bernard Parish Homeowner, March 18, 2006).

Even people who did not live in the affected area were concerned about safety. Suddenly, not only grassroots activists and other concerned citizens were troubled by potential environmental health threats. The hurricane and especially the oil spill amplified concerns. Residents who lived several blocks from the affected area complained of dense oil fumes permeating their homes and making them sick as they tried to clean up their property.

Murphy Oil Company hired the Center for Toxicology and Environmental Health (CTEH) to conduct testing in the spill zone. Despite its name, CTEH is a for-profit company whose clients largely consist of energy corporations that have environmental accidents. According to a report issued by the Louisiana Bucket Brigade, CTEH conducted tests for Texaco in Ecuador after they dumped 18.5 billion gallons of toxic waste into the drinking water of 30,000 rainforest residents (Louisiana Bucket Brigade, 2006). As we shall see in the forthcoming chapter, questions will once again be raised about CTEH's relationship to the oil industry.

CTEH, in its September 21, 2005, report to Murphy Oil, seemed to minimize the harm: "The presence of petroleum from the oil spill in some of the homes poses no additional hazard to homeowners." A month later, in another report dated October 21, 2005, CTEH stated, "[T]he spill should not be expected to present any long term health and safety issues."

On November 8, 2005, the Centers for Disease Control and Prevention's Agency for Toxic Substances and Disease Registry (ATSDR) issued a report warning about short-term exposure. It recommended that people should not return to their homes until they were free of oil and that children and pets should be restricted from contaminated areas. However, the report failed to discuss possible long-term health threats. Much to the confusion of residents, a couple of days later, EPA administrators stated that it was too early to determine when the area would be safe and if it ever would be safe to reoccupy some homes. Despite these concerns, neither the state nor federal government made official restrictions on homeowners moving back into their homes. David Payne, attorney for St. Bernard Parish in conversation with EPA officials, asked if it was safe for residents to return to their homes. He was told it was safe, but when he asked the EPA to put the assurance in writing, the Agency refused (Payne, 2006).

Many residents, along with the Bucket Brigade and the NRDC, believed the sampling being conducted by Murphy Oil's contractor CTEH and the EPA was woefully inadequate. The sampling protocols raised many concerns. Among the concerns was that tests were only conducted for the components of crude oil. Moreover, CTEH only tested surface samples and did not conduct core-sample testing, the protocol for oil spills—especially in porous soils like those in St. Bernard Parish. Furthermore, CTEH contaminated the samples by actually mixing soils from different sites and bagging them together. The EPA was supposed to conduct split sampling: that is, 10 percent of the samples were to be independently analyzed by the EPA. However, the EPA used the same lab as CTEH, and verifying differences in soil samples was difficult, which meant that, ultimately, there was no split sampling. Finally, Murphy Oil allowed CTEH to make the decision about which soil samples to collect, thereby increasing the likelihood that a for-profit company might choose to test areas that appeared cleaned.

The Louisiana Bucket Brigade, leery of the testing protocols, spent $20,000 of its own money to conduct independent tests. These tests found that arsenic, cadmium, and various benzene compounds were detected at levels that exceeded both the EPA's and the LDEQ's acceptable levels. Wilma Subra, a MacArthur "genius" award recipient and a chemist who conducted research on Love Canal for the EPA, reviewed the test results and stated flatly that pregnant women, the elderly, and children should not reenter the affected area. Other than the potentially long-term threat of cancer, people exposed to such contamination, she warned, were at risk of allergic reactions and respiratory illnesses.

Gina Solomon, a senior scientist with the NRDC, faulted the CDC report for the failure to monitor air samples within homes and, with other critics, condemned the failure of CTEH to take core soil samples (Solomon, et al., 2006). In the extremely porous soil of the Parish, oil could easily penetrate the soil and later vaporize in the typically extreme heat. Such vapors could not only be released into the atmosphere from open land but, more important, from land under homes and buildings—the vapor could be released into the closed environment of homes. Since oil has been shown to persist in the soil for decades, the problem of such contamination was long term. Much of the soil in the community is sand, which is a perfect place for oil to hide.

Christine Bisek (2006) is both a toxicologist and an attorney and a lifelong resident of Chalmette. She has dealt with environmental contamination issues for years, working for both public agencies and law firms. She was quite shocked by the protocols employed in response to the spill: "I find that the EPA and LDEQ have gone off to Neverland. It's the Neverland of sampling. I have personally experienced them doing stricter sampling from a spill in a tank yard. As a toxicologist I have had to do stricter sampling in such a case than they are requiring in a residential area. I just find it strange" (Bisek, 2006).

Aside from the protocol concerns, many citizens were disturbed that the sampling was fairly small. Furthermore, no long-term sampling or long-term health monitoring of the residents was planned. This left community members with relatively few test results, depicting at best, a partial and inconclusive data set taken at only one point in time.

Grassroots activists and citizens were also concerned that the EPA allowed LDEQ to set its own allowable levels for toxins, which in many cases are less stringent than those of EPA. For instance, LDEQ has set Risk Evaluation/Corrective Action Program (RECAP) standards for Benzo(a)pyrene—a substance very deleterious to human health with a long half-life—at 0.33 mg/kg, while the EPA has set RECAP cleanup standards at 0.062 mg/kg. Many people felt the EPA should enforce the federal standards to protect the health and safety of St. Bernard Parish residents.

Hugh Kaufman (2006), a former high-level EPA analyst and official, was very critical of the way the EPA conducted itself:

> The EPA has done limited air, water and soil sampling. [However,] we still don't have an adequate mapping of the contamination on the ground from all the sediments that have settled in the area. So we don't know where all the hot spots are that need removal. We're a year later and that mapping still hasn't been done. ... [I]t's money versus people. (Bennett, August 24, 2006)

In response to these, as well as the concerns of other residents in the greater metropolitan New Orleans area and other environmental groups, the NRDC pressed for more testing and monitoring of the region for possible environmental contamination from Hurricane Katrina and the flooding. Mike D. McDaniel, the secretary of the LDEQ, wrote a letter to the press that labeled individuals who called for more precautionary testing "alarmists" and "scaremongers." He claimed that such individuals and groups were making "inaccurate, misleading, and often outrageous claims," which he described as a "steady drumbeat of false alarms" (McDaniel, 2006). This letter infuriated many residents. It reinforced a common perception of LDEQ— based on its checkered, controversial, and political history—as a service arm of the petrochemical industry. Once again, the agency was failing to protect the public from harmful toxins.

Anger, anxiety, and confusion surrounded the manner in which the testing was done. Murphy Oil and the EPA were not forthcoming in reporting the results. Residents complained that they had a right to know the facts as soon as possible. In their view, their lives hung in the balance, and they were in perpetual limbo about whether to return home. The following statements are indicative of the uncertainty created in the absence of information:

If the EPA misled the public about air pollution safety after 9/11, why should we trust what they have to say in the wake of Katrina?

How bad is it really? The Environmental Protection Agency says one thing, the Louisiana Department of Environmental Quality says another, and Murphy is telling me I don't have a problem. What am I supposed to do about everyone saying different things?

Murphy said my property was positive for contamination and should be cleaned while the EPA refuted Murphy and said my property didn't need to be cleaned.

My neighbors and I all live on small lots. When Murphy tells me my property is heavily contaminated and my neighbors on all sides of me are not even those who live down hill—how can we believe them? Does oil pollution not pass through chain link fences? (Attendees, Louisiana Bucket Brigade Community Meeting, March 19, 2006)

Joy and Johnny Lewis lost two homes and one business from the spill and the hurricane. Joy lost her mother, who, like Joe Navis's mother, drowned in the St. Rita's nursing home in St. Bernard Parish. Joy was also disappointed in the response of Murphy Oil:

They were supposed to come here and sanitize it [the property] and spray it and then remove six inches of top soil off the ground and replace that with new top soil. But then they said the oil just stayed on top. That's not true, it penetrated the china that I had and it must have also penetrated the land. (Lewis, March 2006)

Facing class-action lawsuits, Murphy Oil offered early settlements to residents on the condition they waive their right to legal action. The company offered homeowners $20,000 in compensation per home and $2,500 to each resident of the home with the promise that the company would clean the homes and property if they were found to be contaminated. After local attorneys petitioned, a federal judge required Murphy to advise property owners that they should seek the advice of counsel before they waived their rights to sue. While recognizing that the settlement was meager and that there were many future uncertainties, nevertheless, many residents settled, often compelled by extreme hardship and desperation.

Joe Navis and his father were among them. Joe and Sid Navis had just lost Joe's mother in the flooding of St. Rita's nursing home in St. Bernard. Joe and his father faced many financial hardships because of their medical conditions and that of Joe's daughter. Joe and Sid knew that they might stand to gain more in the long-run. In the short-term, however, they were faced with homelessness, disability, and medical bills. Like many elderly residents, they were concerned about potentially long, drawn-out lawsuit that might not be settled in their lifetime. Joe Navis noted, "Well, for my dad at seventy-eight, I think we should settle, you know, because a class-action suit is going to take a long time. But I don't think they are offering much. I think their offer is peanuts" (Navis, J., February 26, 2006).

Defining Ground Zero

The homes of many disaster victims in Seveso, Love Canal, and the Shetland Islands failed, ultimately, to be included in the officially declared contaminated zones (Whiteside, 1979; Levine, 1982: Button, 1995). Many residents of New Orleans feared that their homes, too, might not be included on Murphy Oil's map or the court map of houses eligible for the class action suit. The two maps were not identical. There was considerable controversy over whose homes were included and whose homes were excluded. Joe Navis summarized some of the anxiety:

> It's in all kinds of flux right now. It is ridiculous. And they have drawn a map of the oil spill area and my dad's house just made it on the map, like the map goes just two blocks beyond my dad's street. And there was oil everywhere, as if the oil stopped in a certain place. The oil was throughout the entire parish. It's just a boondoggle. We are lucky to be on the map. (Navis, J., February 26, 2006)

Another resident expressed concern about the apparent arbitrariness of the map:

> The Murphy map has the oil ending right at the corner of our street, as if it just flowed there and stopped. We have no clue if we are eligible for a settlement or not. We too have oil on our fences and on our bricks. (St. Bernard Parish Community Member, March 15, 2006)

Some experts estimate that from 200 to 500 homeowners who claimed damage from the spill were not included in either the Murphy Oil company settlements or the class-action suit. Some residents that waived their rights and accepted Murphy Oil's settlement later complained of their treatment at the hands of the oil company. Some alleged that Murphy Oil originally told them their house was contaminated with oil. Later, after the homeowners signed the agreement, Murphy said there wasn't enough oil to warrant cleaning their houses. Others complained that after they signed the settlement Murphy Oil would not remediate their property even though oil was visible on their land. Ultimately, 2,880 homeowners accepted Murphy Oil's offer.

In January 2006, a federal court certified the class action suit filed by approximately 6,000 plaintiffs. The court chose a class area larger than Murphy Oil's settlement zone map and smaller than that offered by the plaintiffs (*Patrick Joseph Turner, et al. v. Murphy Oil USA, Inc.*). A settlement was reached in January 2006 in which the federal judge ordered Murphy Oil to pay $330 million to the plaintiffs. Damage compensations ranged from a flat $15,000 up to $40 a square foot and $3,375 per occupant. Fifty-five million dollars of the suit was to be set aside to buy or demolish homes (570) in the area nearest the oil refinery and conduct an environmental cleanup. It is important to point out that, within the spill area map established by the court, there were four zones and that in some zones home owners could receive as little as $10 a square foot. Seventy million dollars was also set aside to remedy environmental damage in the overall spill region (Finch, 2007).

After the Settlement

After the settlement was reached, Murphy Oil promised that it would not expand its plant into any of the court-ordered buffer zones that were to be created by the company's purchase of the most heavily contaminated homes. Since the settlement, Murphy has bought 356 homes in the affected area. However, in 2007, Murphy sought permission to add four

250,000-barrel tanks to its plant. One community spokesperson argued against the expansion: "I don't think they should be allowed to add more tanks until they can insure the ones they have are safe" (Rioux, 2007). Among the concerns that residents had about the proposed expansion is that the current oil spill plans, overseen by the EPA, do not require an infrastructure to prevent inundation from catastrophic flooding, such as occurred during hurricane Katrina. Air sampling conducted in June of 2007 near the refinery showed an average benzene reading of 5.83 parts per billion, according to Anne Rolfes (Rioux, 2007). This reading is above the LDEQ's 3.76 parts-per-billion standard for year-long exposure.

Immediately after Katrina, many residents stubbornly refused to sell their homes and insisted on returning. As Anne Rolfes stated, "People want to remain in their homes not only to preserve their neighborhood, but on principle because they don't want to relinquish their land to a refinery. They don't want to envision the community as being relegated to predominately being a refinery." (2007)

The rebuilding of St. Bernard Parish has been extremely slow, even compared to the slow recovery of other New Orleans communities. According to the New Orleans Index, only 41 percent of the pre-Katrina households are actively receiving mail, and public school enrollment is about 40 percent of the pre-Katrina total (Bookings 2008). The same report states that the private school (primarily parochial) enrollment has regained only 26 percent of its pre-storm enrollment. Childcare centers are practically nonexistent. The sewer system has not been rebuilt. FEMA trailers cover acres of land. Few hospitals have reopened. Few food stores and gas stations serve the area, and residents often drive great distances for either service. Entire neighborhoods are gone. Houses that had been occupied continuously by several generations of one family are now vacant.

When I began interviewing residents in the initial months after the storm, most said they would return to the neighborhoods where they had lived most, if not all, of their lives. On subsequent visits, I could find fewer and fewer of these residents. Those who remained were doubtful they could hold on. Many expressed difficulty paying a mortgage on an unoccupied, damaged home while paying another mortgage or rent else-where. Many also found that the long commute from their second home back to their former neighborhood was too time consuming. Finally, each time I returned to the neighborhoods, I was lucky to find, at most, one household working on their home. Typically, homeowners said that they intended to sell. One resident, as he pointed around the block to all the empty or demolished homes, said there was no point in returning. One-by-one neighbors had not returned, and the once-cherished neighborhood was now but a memory.

As fewer people remain behind, the challenge for people like Anne Rolfes and the Louisiana Bucket Brigade and other grassroots organization to organize has greatly increased. As Rolfes observed, even immediately after the storm and spill, many people concerned about harmful environmental effects did not return to live, only to visit and continue to organize. Now that many people have followed, leaving for various reasons—environmental concerns, jobs, affordable housing, schools, active neighborhoods, or a combination of factors—the challenge to organize is greater and the future of the Parish remains uncertain. Although the resettlement has been painfully slow over the last five years, the community has begun rebuilding.

While the residents of St. Bernard Parish have made great strides in their efforts, the Deepwater spill has brought new challenges and uncertainty to the community. The statements made by St. Bernard residents in a recent meeting with BP officials and the U. S. Coast Guard well illustrate the new obstacles that now confront residents. Cheryl Parnhard's eloquent statement expressed the sentiments of many:

> Do you see this community? This is all who came back. These are the people who lost their photos, friends, and our homes. Our hearts and souls have been ripped out of us. Don't stand there and watch it be destroyed and tell us you are doing all you can, because you are not. We have come too far as a city and community. We have cashed in 401ks and our lives have come back to this town. You are not doing enough. (West, June 1, 2010)

Others with whom I spoke expressed similar frustrations. Imagine, as one man said, yourself in their place:

> It is bad enough that we had to suffer through Katrina and the Murphy Oil spill and the enormous difficulty these last few years to rebuild our community, but now we have to put up with the pain and uncertainty of BP's negligence. We are struggling to survive in the face of one disaster after another.
>
> When I think of the pain, sweat, and tears my family and neighbors have had to endure to get this place back up on its feet it doesn't

seem fair that because of corporate greed and government's failure to protect us that we have to suffer through this. Do you people have any idea what we are going through? Can you imagine yourself in our place? (Simpson, June 13, 2010)

5. Knowledge Withheld

People are still struggling with fractured support systems, loss of property, sense of self, income, community and loved ones, stigma, unstable living situations, transportation problems and difficulty acquiring quality health care and childcare. It is unacceptable that families must endure uncertainty and concerns regarding possible short and long-term effects of on-going exposure to elevated levels of formaldehyde in addition to the daily anxieties and stresses of displacement.

Heidi Sinclair, MD, Medical Director, Baton Rouge
Children's Health Project (April 1, 2008)

Verna leaned over to pick up her crying four-year-old daughter who was coughing continuously. I put down my pen and turned off the tape recorder. Verna told me that in the last few days since she and her mother moved into their new FEMA trailer with her two daughters, Lily, her younger daughter, had begun to have teary red eyes and a "God awful" cough that was worsening. Her older daughter, Carrie, seven, was also showing signs of respiratory distress. According to Verna, neither of her daughters had any of these symptoms before moving into the trailer. She asked me if it was possible that it had anything to do with the trailer. I told her it was possible, that there might be some kind of off-venting causing the problem, but that I was not sure. This was my introduction to what would become the FEMA trailer syndrome and

FEMA Trailers.
Photograph © Gregory Button.

the beginning of Verna's long struggle to learn why her daughters were having health problems. At that time, no one could have imagined the bizarre twists and turns this story would take or the plots and subplots that would emerge over the coming three years.

Verna, her mom, Ella, and I were not aware at that time that hundreds of other families were beginning to experience similar problems. It would be weeks and months before the full story emerged as clues to what was wrong with many of the FEMA trailers unfolded slowly. Families living in the trailers struggled with the uncertainty of their illnesses and the denial and eventual coverup by federal officials of the true cause of the problem. Eventually, the facts surrounding these mysterious illnesses would become another significant chapter in the Bush administration's failure to respond adequately to Hurricane Katrina—this time it would not only implicate FEMA, but the CDC and its sister agency ATSDR as well. These revelations ended up taking an inexcusable amount of time.

Not long after conducting this interview with Verna, I visited with sixty-eight-year-old Jimmy, a retired shrimper and jack-of-all-trades. Jimmy and his wife, Louisa, had just returned to St. Bernard Parish, Louisiana, from Reston, Virginia, where they had been living with their daughter. They were excited about returning to the Parish and beginning their life anew. It would be a while before they could again inhabit their home, which sat only a few yards behind their trailer. Their house had suffered tremendous damage from the flood of Hurricane Katrina, and very few of their personal belongings remained. The water line nearly reached the rafters in their one-story bungalow, and only a sodden, moldy couch and an armchair remained in their living room. In the small den behind their kitchen were the only other signs of their thirty-five years in the house. On the wall hung two severely water-damaged photos that Jimmy had taken of his young family years before. The long-standing flood water had erased the details on the faces of the happy family posed on their boat. Next to them hung a photo of Jimmy in his Air Force uniform, taken before he shipped out to Vietnam. That was pretty much all that remained of their former life. They would take their few visitors into the back room, show them the two faded photos, and proudly talk about their happy life in their home.

They had been in the trailer only a few days, but on the tiny kitchen table were more recent family photos heralding their new life and their hope for a return to normal. Before long, unfortunately, their hope turned to tragedy. Jimmy, a lung cancer survivor of a few years, and his wife Louisa, who suffered from leukemia, began to experience hacking coughs, rashes, shortness of breath, and prolonged headaches. Jimmy was certain his troubles were attributable to his bout with lung cancer; Louisa thought perhaps her immune system had been weakened from chemotherapy and the stress of the hurricane. After multiple visits to physicians and endless tests, however, their certainty waned. They were puzzled by their discomfort. Inexplicably, they eventually attributed their problems to the trailer, especially after learning of other neighbors having similar symptoms. They then did the unthinkable; rather than move back with their daughter and leave the community in which they were born and raised, they moved into their shell of a home. They covered walls in one room with plastic, installed fans in opposite windows for ventilation, and laid a makeshift plywood floor. Jimmy secured the trailer from neighborhood kids with a padlock. In time, however, they were forced to move back to Virginia.

Variations on this theme were repeated endlessly. Frequent headaches, nosebleeds, and coughs were commonplace. On my third visit with Verna, she reported that she had to rush her four-year-old daughter to the emergency ward in the middle of the night because of a severe asthma attack. Other people cited lingering coughs, sinus infections, and dizziness. Some reported bronchitis and pneumonia. Jimmy's eighty-year-old neighbor complained of a "flu that would not leave him alone" (Johnson, April 14, 2006). One young mother explained that she often could not sleep through the night. She constantly had to get up to care for her three children who woke up with hacking coughs and red eyes.

Following Hurricane Katrina, Lindsay Huckabee of Kiln, Mississippi, and her husband and four children (and expecting another) moved into a FEMA trailer on December 14, 2005. As with many other families, hers had difficulties almost from the beginning. As others, too, had reported, first there was a strong odor and then "our whole family began to have sinus problems; our eyes would burn and water, and our throats were constantly sore. We seemed to catch every cold and virus going around, but we couldn't get rid of the illnesses. Three of our children began having severe nose bleeds. I began having migraine headaches and pre-term labor" (U. S. House, Testimony of Lindsay Huckabee, July 19, 2007).

Her four-year-old daughter, Lelah, suffered the most. Over the next year and a half, the child had countless ear infections, numerous nosebleeds, and several turns of pneumonia, for which she was twice hospitalized. Lindsay later testified before a Congressional hearing that FEMA was treating trailer occupants as "lab rats."

Forty-six-year-old James D. Harris Jr., his wife, Aretha, and his family lived with his parents and his brother's family of four, in one room of his parents' house that was left untouched by Hurricane Katrina. Eight months later, in April 2006, he and his wife finally received a FEMA trailer. Entering it for the first time, they "noticed a pungent and overpowering odor that permeated" the structure. When they complained to the FEMA maintenance call center, they were simply told they needed to air out the trailer. FEMA took no further action. Both Harris and his son experienced ill effects from the off-venting of the formaldehyde (U. S. House, Testimony of James Harris, July 19, 2007).

The story of Paul Stewart and his family further illustrates the difficulties trailer residents experienced. FEMA delivered a camper to the Stewart family in December 2005. Their camper had what he called "a strong 'new' smell." They took the precaution of airing it out for a few days before they moved in. No sooner had they moved, however, than trouble began:

> The first night we stayed in the camper my wife woke up several times with difficulty breathing and a runny nose. She got up once and turned on the lights to discover that her runny nose was in fact, a bloody nose. This scared the hell out of us; we didn't know what was causing the bloody nose, or breathing issues and I was beginning to show symptoms of my own, which included, burning eyes, scratchy throat, coughing and runny nose. The symptoms continued for weeks and then months. (U. S. House, Testimony of Paul Stewart, July 19, 2007)

Others and I soon learned that social workers, visiting nurses, and community volunteers were seeing the same problem repeatedly. Soon the common knowledge along the Gulf Coast was that the trailers FEMA had provided the evacuees of Hurricanes Katrina and Rita had some kind of toxic off-venting. Some people began spending all their waking hours outside their trailers, enduring the heat and humidity of the Gulf Coast. Others erected tents and makeshift lean-tos, using their trailers only for bathroom and kitchen facilities. After spending time in the trailers, I was also beginning to experience eye and throat irritation. Some social service people were beginning to wear paper dust masks when making their calls on trailer families. Social workers and nurses reported that some elderly trailer residents were hospitalized for acute respiratory problems;

many of these people were dying. That people expected an explanation from FEMA for these inexplicable symptoms was understandable.

Rumors began to circulate as to the cause of the problem. In the wake of the confusion, concern and uncertainty abounded. People did not know what to think, but they began to demand answers. Rumors of people dying in their trailers or the emergency room gained momentum. Suddenly, yet another post-Katrina problem demanded an urgent response, an explanation, a solution. People were at first reluctant to complain to FEMA because they were afraid they would be evicted from their trailers. If they complained, they feared their application for disaster relief assistance would be stalled or stopped altogether. "Where would my family and I go if FEMA kicked us out of our trailer? There are no apartments or homes for rent in the area and we can't afford to pay for a place ourselves," stated one man (St. Bernard Parish Resident, April 15, 2006). Many residents despaired; knowing what to do became impossible.

As would later be learned, the problem began when, with hundreds of thousands of people homeless, FEMA and the Department of Homeland Security (DHS) decided to house an estimated 12,000 families, 275,000 people, in trailers. Demand exceeded supply. FEMA rushed to sign contracts with Gulf Stream Coach, Fleetwood Enterprises, Monaco Coach, and other trailer manufacturers. Later, a DHS employee would testify that the contract lacked proper health and safety specifications. Others would later testify that building materials for the trailers were scarce and that eventually manufacturers began importing wood composite board from foreign countries that had high formaldehyde content. For this story to unfold, required the initiative of trailer occupants, like Paul Stewart and other FEMA occupants, along with Becky Gillette, someone whose own home was flooded by Hurricane Katrina. Gillette became the Sierra Club formaldehyde coordinator.

Like others, Paul Stewart had complained unsuccessfully to FEMA about his family's health problems. Initially, FEMA told him to air out the trailer. When he told FEMA that he had tried that remedy, but to no avail, "FEMA's response was hollow and degrading, they just told us there was nothing else they could do." At one point Stewart said that, "FEMA responded that I should 'be happy to have what I have' and told me that, 'we do not have any more campers available so you will have to make do with what you have'" (U.S. House, Testimony of Paul Stewart, July 19, 2007). Others received similar responses.

Lindsay Huckabee later testified that FEMA, rather than addressing the concerns of trailer residents, "instead [made]people ... feel that they were being too prickly, or looking to blame someone else for simple colds and normal problems" (U.S. House, Testimony of Lindsay Huckabee, July 19, 2007). FEMA's usual response was the recommendation to open the doors and windows of the trailer. Then it recommended spending more time outside, cleaning up mold, controlling moisture, and avoiding smoking in the trailer. Later, an internal FEMA report would find that FEMA policies regarding response to such complaints "were unclear and inconsistent" (Office of the Inspector General, June, 2009: 20).

Paul Stewart eventually decided to take matters into his own hands. He decided to test his own camper and ordered a formaldehyde testing kit from a Florida company, America Chemical Sensors. When he described the problems he and his wife were experiencing, the company responded that the symptoms clearly indicated formaldehyde poisoning. They sent him the test kit free of charge. The lab results indicated that the formaldehyde levels were twice that of the EPA's safety limits. Despite the test results, FEMA was not interested in helping him. Investigating on the internet, he discovered that OSHA, forty-three days after the hurricane, began conducting tests for dangerous chemicals including in a FEMA resettlement camp in Pass Christian, Mississippi. OSHA discovered the outside ambient air tested very high for off-venting of formaldehyde that was being released from inside the trailers. Eventually, he contacted a local television station, WLOX in Biloxi, Mississippi. After his story appeared on the ten o'clock news, the Sierra Club contacted him. He joined forces with the Sierra Club, and together they tested sixty-eight campers: sixty had dangerous levels of formaldehyde.

In May 2006, the Sierra Club announced the results: most of the FEMA trailers had elevated levels of formaldehyde. According to a House Congressional report (CITE), typically less than ten parts per billion (ppb) is reported in outdoor, ambient air, and indoor concentrations range from ten to thirty ppb (Centers for Disease Control, November 10, 2009). The Sierra Club tests showed that in 83 percent of the fifty-two trailers tested formaldehyde levels of over 100 ppb. Once these results were made public, the media picked up on the story, first locally and then nationally. Media floodgates opened, and Hurricane Katrina evacuees were no longer the only ones asking questions. Despite the widespread press accounts, FEMA still denied any problem.

Becky Gillette persisted in trying to obtain more information about formaldehyde levels in the trailers. More than a year after Hurricane Katrina came ashore, the EPA began testing. Gillette was anxious to obtain whatever information she could from the tests, because, as she later testified, she welcomed the "more expensive, extensive" testing. However, after repeated attempts, she was unable to obtain results. FEMA claimed the results were delayed because the tests were sent to ATSDR for evaluation. As we will learn later, ATSDR's involvement generated considerable controversy. Finally, Gillette sent a Freedom of Information Act (FOIA) request to FEMA to obtain the EPA results followed by repeated emails to ATSDR—all to no avail. In one of her emails to James Durant, an environmental health scientist, she expressed her frustration and asked for advice. Here is Durant's response, according to Gillette's testimony:

> I am sorry it took so long to get back to you. My supervisor and I have been trying to track down who in CDC/ATSDR has been heading up this issue. This was not as straight forward as we thought it would be. We have found the person heading this up, unfortunately she is out of office. Hopefully, we will be able to get an answer to you on what is going on with the formaldehyde soon. (U.S. House, Testimony of Becky Gillette, April 1, 2008)

After a couple of weeks, she emailed Durant again, making further inquiries and asking if he had any information. Again, his response:

> So you are telling me that no one has contacted you regarding formaldehyde at all? When you contacted me, we attempted to have the person who is heading this up contact you. It was my understanding you would be contacted. I will flag the issue and try to get someone to contact you that knows what is going on. (U.S. House, Testimony of Becky Gillette, April 1, 2008)

At the time, most people did not know that OSHA had conducted its own tests beginning in October 2005 and ending in January 2006. In the Gulf Coast region, it conducted over 100 formaldehyde tests including in FEMA trailers. Since the agency is concerned with worker, rather than residential, health and safety, the tests monitored the workplace. Some of the FEMA trailers tested very high. For instance, in Mississippi, formaldehyde readings in some trailers were as high as 590 ppb. These results were then sent to the contractors managing the sites of the trailers. The contractors, however, did not forward the test results to FEMA. In March 2006, OSHA faxed the results to FEMA (Office of the Inspector General,

June, 2009: 15). The ATSDR's health consultation on the matter "FEMA Study: Ventilating Trailers Can Significantly Reduce Formaldehyde Emission Levels," was not released until May 2007. Below is an excerpt from the report:

> The baseline for concentrations of formaldehyde in the units averaged 1.2 ppm at the beginning of the test. …According to the evaluation report provided to FEMA by ATSDR, the average concentration of formaldehyde per day in units using open window ventilation dropped below 0.3 ppm after four days of ventilation and remained low for the rest of the test period. The level for health concerns for sensitive individuals was referenced by ATSDR at 0.3 ppm and above. (Agency for Toxic Substances and Disease Registry, 2006: 19)

The results startled Gillette and others. First, ventilation does not work in the summer, as Gillette had already pointed out, because of the high heat and humidity. The intake of high humidity levels also exacerbates the off-venting levels. Moreover, as Gillette astutely observed, 1.2 ppm is considered extremely high. She also expressed concern that ATSDR stated in their report that levels below 0.3 ppm posed no health concerns. As she pointed out, "At that level, most people experience extreme distress. ATSDR's own standards are "many magnitudes lower at 0.4 ppm for 1-14 days and far longer for long-term exposure" (U. S. House, Testimony of Becky Gillette, April 1, 2008). This last point would become critical, as we will see a little later, when FEMA urged ATSDR to restrict their analysis to only short-term exposure levels. Levels limited to short-term exposure ignored the fact the trailer residents were living, and would live, in the trailers for extended periods. Becky Gillette responded: "What ATSDR did was criminal negligence covering up this problem when the health and lives of tens of thousands of Americans was at stake" (U. S. House, Testimony of Becky Gillette, April 1, 2008). The controversy over formaldehyde was about to burst, dragging the CDC and ATSDR center stage along with FEMA, the EPA, and the trailer manufacturers.

Within FEMA, some officials were concerned about establishing safe levels of exposure to formaldehyde. While there were established exposure standards in the workplace (OSHA's allowable time-weighted average was 750 ppb for eight hours), no federal standards had been established for residences. Recommendations from various federal agencies were not uniform and often conflicted, with the exception of HUD, which only had standards for mobile homes, not travel trailers or park trailers. In 1984, HUD set a target of keeping formaldehyde exposures in mobile homes below 400 ppb, a standard that exceeded what many other experts would recommend. Amid this confusion, some FEMA officials felt they needed guidance from another federal agency. Local FEMA officials became concerned about the apparently high levels of formaldehyde in some of the trailers and wanted the agency to do something to alleviate the exposure levels. However, their hands were tied because they could not make policy decisions. Upper-level FEMA officials were reluctant to act when no other federal agency had set thresholds for residential formaldehyde levels. They and other FEMA employees were not aware of a cautionary letter from a top CDC/ATSDR official, withheld by a FEMA attorney. Thus, they not only failed to react properly, they actually issued statements to Congress, the media, and the public that they were unaware of any significant health hazards in the trailers (Office of the Inspector General, June, 2009).

Responsibility and Litigation

In August 2006, a major news network made a Freedom of Information Act Request (FOIA) for the formaldehyde test findings. The FEMA legal department responded that the data would not be released because it was initiated in response to a legal issue. "[T]he testing was undertaken because FEMA was sued. ... The testing is covered under the following exception to FOIA #5 and has been prepared in anticipation of litigation and is covered under deliberative process privilege, the attorney work product privilege and the attorney client privilege" (Office of the Inspector General, June, 2009: 37).

This response is highly problematic since all evidence indicates that the testing was proposed and begun before any indications of potential litigation. The response of the acting assistant administrator, Disaster Assistance Directorate, and the response of the recovery staff served to underscore this. "For the record, we initiated this testing before we were sued. [And wrote to his staff] is this right? I was not aware of any litigation when you first proposed engaging the EPA to test. The staff Recovery responded: I don't know if we were aware of the litigation when we began working with EPA (it certainly wasn't the driving factor)" (Office of the Inspector General, June, 2009: 37). Certainly the rationale for denying the FOIA request does not concur with official statements made by FEMA officials about why they requested the EPA tests. Here is a statement from the FEMA news desk: "The agency has specially asked for and received from the Environmental Protection Agency (EPA) an air monitoring and sampling plan that is intended to validate scientifically, methods that can be used to reduce the presence of formaldehyde in trailers" (Office of the Inspector General, June, 2009: 38).The FEMA Public Affairs Office made a similar statement: "The purpose of this study is to provide scientific support for methods that can be used to reduce the presence of formaldehyde in trailers. Specifically, the results will be used to identify activities we can take and that we can instruct the occupants to take to lower the levels of formaldehyde" (Office of the Inspector General, June, 2009: 38). In response to the continuing uncertainty and public health concerns, Congress took action. In July 2007, the House Committee on Oversight and Government Reform held a hearing on FEMA's failure to "respond adequately to reports of dangerous levels of formaldehyde" in FEMA trailers provided to Hurricane Katrina and Rita evacuees. The hearing revealed that as early as March 2006, FEMA recognized the dangers of formaldehyde in the trailers. Even though the FEMA field staff urged immediate corrective action, FEMA's Office of General Counsel prevented any action stating, "Should they indicate some problem, the clock is running on our duty to respond." During the hearing, FEMA's administrative director stated that FEMA "could have moved faster" (U.S. Congress, July 9, 2008).

Among those who testified at the hearing was Scott Needle, representing the American Academy of Pediatrics. Scott is a general pediatrician in private practice in Bay St. Louis, Mississippi, who personally witnessed the deleterious effects of high levels of formaldehyde in the trailers. He testified that, beginning in the spring of 2006, a pattern emerged of children having repeated visits to him for treatment of recurrent sinus infections, ear infections, colds, and other respiratory symptoms. Two underlying factors emerged from Scott's questioning of the parents: They lived in FEMA trailers and the symptoms started shortly after they moved into the trailers. On behalf of the Academy, Scott expressed serious concerns about the short- and long-term health effects of continued exposure to formaldehyde. Moreover, the Academy urged federal health agencies to conduct systematic and rigorous studies of the health issues and initiate steps to safeguard the children.

Because of the hearing, Chairman Henry A. Waxman initiated an investigation into what the manufacturers knew about formaldehyde levels in their trailers. A year later, in July 2008, the House Oversight

Committee issued a majority report on the trailer manufacturers. Over 9,000 pages of documents revealed that even though the manufacturers had also received complaints from evacuees and even after the manufacturers conducted tests on their trailers, they did not inform FEMA of elevated levels of formaldehyde. The largest supplier of trailers was Gulf Stream, which also conducted the most extensive testing of trailers. In every occupied trailer that was tested, the company found formaldehyde levels at or above 100 ppb according to the Oversight Committee executive summary. Two of the trailers tested at 600 ppb. Higher levels, some as high as 4,000 ppb, were discovered in unoccupied trailers. Forest River, another supplier of trailers, hired a contractor to test its trailers and reported that one trailer had levels of 15,000 ppb. The company was advised by the contractor, according to the same Oversight report, to "post signs … stating 'hazardous-do not enter'." To put this information into perspective, many health agencies and organizations believe acute, harmful health effects can be experienced at levels above 100 ppb. Levels at 100 ppb and higher are considered too high by the CDC, the Environmental Protection Agency, the World Health Organization, the Consumer Product Safety Commission, and the National Institute for Occupational Safety and Health.

According to the Oversight Committee majority report, Gulf Stream's own internal records demonstrated that the manufacturer considered the issue of formaldehyde in their trailers "a public relations and legal problem, not a public health threat." The company even admitted to the Committee that the testing of the trailers was done not to assess health threats but in anticipation of litigation. Gulf Stream, the report stated, failed to inform the trailer residents of the possibility of exposure to hazardous levels of formaldehyde. Furthermore, the manufacture did not advise FEMA against placing evacuees in any of the unutilized trailers. The Committee report went on to state that Gulf Stream told CNN: "We are not aware of any complaints of illness from many of our customers of Cavalier travel trailers over the years, including travel trailers provided under our contracts with FEMA!"

FEMA evacuees were not the only people potentially exposed to adverse health effects. In the process of their investigation, the Committee interviewed former Gulf Stream employees who complained of health effects while constructing the trailers. These effects included familiar symptoms also reported by FEMA trailer occupants: nose bleeds, shortness of breath, dizziness, ear bleeding, and sinus infections. Here is what one employee reported about the wood material used to construct the trailers:

> One of the employees reported that the wood materials Gulf Stream used in the production of the travel trailers for FEMA had a "foul odor" and the odor "got worse" as "more came." He explained that he realized the odor was coming from the wood materials used in the assembly of the travel trailers because he "would take two pieces of that and lift one up" and that "it was just overwhelming the smell that would come from that."

Another employee told the Committee that, at the completion of trailer construction, the trailers had a very strong odor. She went on to say, "everybody was aware of that odor. It wasn't something you had to report. Everybody smelled it."

The FEMA trailer scandal was not, by any means, put to rest by the Waxman hearings. The House Committee on Science and Technology held its own hearing in May 2008. The Subcommittee on Oversight and Investigations staff read hundreds of hours of interviews and thousands of pages of documents in preparation for the hearing. Ultimately, the Subcommittee uncovered plots and subplots to this story that had not yet been discovered. Chairmen Bart Gordon (D-TN) of the House Committee on Science and Technology gave an opening salvo that lashed out at the ATSDR. In his statement, he outlined what he described as ATSDR's failures. When, for instance, the agency began working with EPA and FEMA in developing test protocols for investigating the formaldehyde levels, the top, most qualified personnel were not assigned to the task. Instead, the task was assigned to emergency response staff that had no training in toxicology. Furthermore, their report went to the director of the Office of Terrorism, Preparedness and Emergency Response (OPTER) for review rather than through appropriate channels. As we learned from Becky Gillette's account, the report recommended ventilation and seriously downplayed any adverse health risks and ignored any long-term health risks.

Gordon asserted that the wrong people conducted the tests, the wrong people wrote the report, and the wrong people reviewed the report. He further alleged that there was "managerial collapse" at ATSDR when the weakness of the report was made known to the director who then did nothing to rectify the situation. What emerged during the hearing was that the one person that should have been in charge of the research and made the shortcomings of the findings known to the director was made a scapegoat, labeled a whistle blower, and then demoted. Brad Miller, chairmen of the Investigations and Oversight Subcommittee went even further: "The evidence FEMA ignored, hid and manipulated government research on the potential impact of long-term exposure to Formaldehyde by Katrina victims now living in trailers is hard to ignore" (U.S. Congress. House, April 1, 2008e).

The Subcommittee hearings were critical of Howard Frumkin's role as the director of the CDC's ATSDR. As will be illustrated below, the Subcommittee's report unveils how Frumkin and other agency personnel mishandled the research and its shortcomings and attempted to place the blame on whistleblower Christopher De Rosa, an internationally recognized expert with sterling credentials and the head of the agency's division of Toxicology and Environmental Medicine. The hearing demonstrated, according to Chairmen Brad Miller, how "the agency issued a scientifically flawed report" and then, when made aware of its flaws, failed to correct the report (U.S. Congress. House, April 1, 2008e). The questions raised in the hearing regarding ATSDR's conduct resulted in a later hearing that focused on the agency's role in other cases.

Dr. Christopher De Rosa worked as a government employee for twenty-eight years before the controversy over the FEMA trailers arose. While he now serves as the assistant director for toxicology at the National Center for Environmental Health/ATSDR/CDC, he had previously served for sixteen years in a much more prominent position as director of the Division of Toxicology and Environmental Medicine at ATSDR.

Dr. De Rosa testified before the Committee that in December 2006 two members of the division's Emergency Response Team (ERT) were requested by the EPA to evaluate its sampling data of formaldehyde in the FEMA trailers. However, FEMA's attorney, Patrick Edward Preston—the same attorney who had earlier cautioned FEMA employees from immediate action on the formaldehyde problem—directed these team members not to share the evaluation through the normal division review and approval channels. Thus, evaluations did not come to De Rosa's attention for review. Instead, the ATSDR employees were instructed to submit their evaluation for review to the director of the Office for Preparedness, Terrorism and Emergency Response (OPTER). These actions, De Rosa stressed, were taken without his knowledge.

Dr. Frumkin and another employee, Dr. Sink, reviewed and commented on the team's evaluation. Unbeknownst to Dr. De Rosa, the health consultation was then forwarded to FEMA. De Rosa was provided a copy at this time. After he reviewed the consultation report, according to his testimony, he "immediately" called and emailed Frumkin regarding what he viewed as the limitations of the report. His principal concern was that the report failed to state that formaldehyde is a carcinogen and to address the long-term health concerns.

Although he had no formal involvement in the FEMA consultation until June 2007, he "repeatedly cautioned" Frumkin and other senior staffers about his concerns:

> For example on June 1, 2007, I wrote Dr. Frumkin outlining my concerns in response to a request from FEMA to identify "safe levels of formaldehyde exposure." I cautioned that since formaldehyde is a carcinogen, it is a matter of U.S. Government science policy, that there is technically, no "safe level" of exposure. I also wrote that in 1995, the World Health Organization's (WHO), International Agency for Research on Carcinogens (IRAC) had classified formaldehyde as "probably carcinogenic for humans" while EPA had determined that formaldehyde is a "probable carcinogen." (U.S. Congress. House, April 1, 2008c)

De Rosa further cautioned, among other things, that the ATSDR's Health Guidance Values for short-term, intermediate, and long-term exposure be used in assessing the formaldehyde hazards. He went on to note—and this is of particular importance—that, to his knowledge, this was the third time FEMA had approached ATSDR requesting health guidance on formaldehyde exposure in the trailers and requested that the agency "restrict our evaluation to short-term eexposure" (U.S. Congress. House, April 1, 2008c).

De Rosa then went on to state that the first time the Agency was approached, FEMA requested him to review a draft of a statement it had drawn up that covered only short-term health information. His response, according to his testimony, was that he indicated to FEMA that the draft neglected to discuss long-term exposures and that this failure could be misleading. If this account is true, then it is interesting that, for whatever internal ATSDR justification, FEMA and the EPA essentially circumvented De Rosa's authority and expertise and somehow managed to have the ERT team conduct the evaluation. Moreover, someone within ATSDR made the additional decision to have the report forwarded not to De Rosa, the division director, but directly to Frumkin, the Agency director.

Despite his cautionary comments, De Rosa, was not kept in the loop as the formaldehyde issue evolved. However, he asserts that he continued to request, without success, that

> health interventions be pursued to address the clinical manifesta-
> tions of acute formaldehyde toxicity presented in clinical settings
> by residents of the FEMA trailers. I stated that such clinical signs
> were a harbinger of a pending public health catastrophe: that may be
> trans-generational in its impact. I stressed the importance of alert-
> ing trailer residents to the potential reproductive, developmental
> and carcinogenic effects of formaldehyde exposure. (U.S. Congress.
> House, April 1, 2008c)

He then claimed that the only response he received was that such matters should not be discussed in emails because they could be *misinter-preted*. If true, then it would appear someone in the Agency did not want a paper trail.

Despite De Rosa's heroic efforts, Frumkin eventually wrote an annual review that stated De Rosa was not considered a team player. After sixteen years of sterling reviews of both him and his division, he was demoted from division director. Ironically, Frumkin gave De Rosa the damning written evaluation just a few minutes before an award ceremony in Capri, Italy, at which De Rosa would receive the prestigious Ramazzini Award, making him one of only 180 elected fellows world-wide in the Collegium Ramazzini. Fellows of the Collegium are consid-ered among the most eminent and prestigious researchers in environ-mental health science.

Not until June 2009, almost four years after Hurricane Katrina came ashore, did the inspector general of DHS issue a redacted report on FEMA's handling of the formaldehyde issue in the trailers (Office of the Inspector General, June, 2009). The report's findings confirm many of FEMA's shortcomings in the trailer controversy. The report begins by stating that, "in our opinion, FEMA did not take sufficiently prompt and effective action to determine the extent of the formaldehyde problem in the emergency housing units once they were aware that such a problem might exist" (Office of the Inspector General, June, 2009:1). It criticizes FEMA for letting almost a year go by in its working with other agencies before determining an effective method to reduce formaldehyde levels in unoccupied trailers and discovering a method (ventilation) that was widely known even before the study was conducted!

Furthermore, the report criticizes FEMA for not addressing the issue of occupied trailers and not fully disclosing the test results. The report continues to state that even though FEMA then arranged with the CDC for a study of occupied trailers, FEMA "caused delays that blocked the study's progress on two occasions." In short, the inspector general's report states that the Agency "did not display a sense of urgency" in responding to a "significant health risk" (Office of the Inspector General, June, 2009:1). It also faults FEMA with not having a formal policy or procedure to deal with the complaints of trailer occupants, resulting in the confusing and inconsistent handling of these complaints. This and other failures resulted in increased and unnecessary uncertainty that exacerbated the suffering of the trailer residents. Because of the failure to have a consistent and formalized policy, the report states, "the needs of occupants reluctant to complain would not surface" (Office of the Inspector General, June, 2009:1). The report also points out that FEMA's failure to have a detailed study of the formaldehyde levels in the trailers prevented FEMA from replacing toxic trailers with ones that were less problematic.

Not the least of the report's criticisms was that information about the extent of the formaldehyde problem was not available until more than two years after the trailers were distributed. Furthermore, the inspector general's report states quite emphatically that attempts by FEMA officials to take effective action and gain the necessary information needed to respond to the problem did not happen until there was widespread media attention to the problem. It also takes FEMA to task for causing delays and issuing a letter of authorization and failing to have an existing agreement with CDC thereby causing a six-month delay.

FEMA then made matters worse, according to the report. Just as the CDC was about to undertake the study, the Agency's concern about not having in place a public relations strategy to deal with the media, Congress, and trailer residents once the final test results were to be announced caused FEMA to have the CDC cancel the subcontractor's study contract until a communications strategy could be designed, rather than working on such a strategy while the tests were being conducted. This caused an additional two-month delay. Finally, one of the major conclusions of the report is that when the study was concluded in January 2008, it underestimated the extent of formaldehyde exposure because, due to delays, by the time the tests were conducted the trailers were already two years old and the tests ended up being conducted during the winter at a time when formaldehyde off-venting was at its lowest levels.

Almost five years have passed since Hurricanes Katrina and Rita wrecked the lives of tens of thousands of Gulf Coast residents. Thousands of hurricane evacuees were forced to live in trailers that have, at the very least, added an additional stressful level of uncertainty to their lives. As a result, many evacuees have incurred deleterious health effects, short-term and possibly long-term. Verna, her mother, and daughters are now living in a small apartment on the outskirts of New Orleans. They are trying to put their ordeal behind them. Her youngest daughter, Lily, still suffers from shortness of breath and has occasional headaches, which keep her home from school. Verna and her mom, Ella, still worry, like many other former trailer occupants, about the possibility of even long-term health effects. "Since we lived in the trailer so long I am worried that I, my mom or one of my daughters could eventually develop cancer or Lily or Carrie might pass on something to their children when they have kids of their own. Who knows? It's not fair we have to live with this uncertainty the rest of our lives" (Johnson, April 14, 2006).

The story does not end here. The controversy surrounding the trailers would resurface yet again, as we shall see in the final chapter.

6. "What We Don't Know Can't Hurt You"

In Eastern Tennessee, in the early morning of December 22, 2008, a waxing moon was reflecting in the chilly, placid, waters of the Emory River. Suddenly, a fly ash impoundment of the Tennessee Valley Authority's Kingston Fossil Fuel Plant collapsed. The force was almost indescribable, sending 5.4 million cubic yards of toxic fly ash into the river and along its shores. Within minutes, the cascading ash spill covered a major portion of a highway, a rail line, and almost four hundred acres of land. In the process, it destroyed three homes, damaged forty-two pieces of property, and affected several hundred residents. When the explosive force subsided, the fly ash deposits in the river were as deep as thirty feet and as high as sixty feet on land. Smaller *ash burgs* ten to twenty feet high were scattered throughout the moonlit landscape making it appear more like a distant planet than the idyllic landscape that existed only moments before. The surrounding countryside seemed transformed in geologic time rather than in a few horrific minutes. Rick Cantrell, who lives in a trailer above the river, compared it to an earthquake and said that the torrent of ash sounded like a freight train rushing through the woods. The changes the ash spill would bring to the lives of local inhabitants would be equally dramatic, and the uncertainties that would ensue as incomprehensible as the changes in the land. Most of the nearby residents, who had already gone to sleep, would awaken in their beds a long way from home, hearing the wash of helicopters circling overhead.

Chris Copeland and his wife were not asleep when the disaster occurred. As they were getting ready for bed, they heard an "awful roaring sound" and the snapping of docks along the river's edge. When they

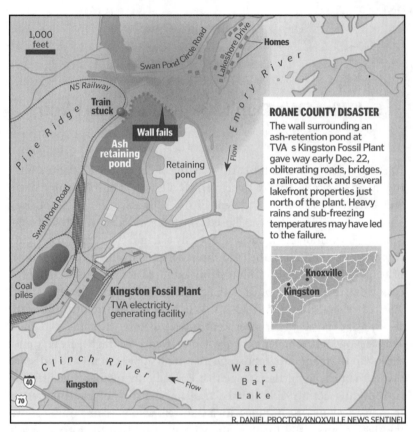

Map of TVA Ash Spill.
Photograph © R. Daniel Patton/*Knoxville News Sentinel.*

peered out their windows, they saw a river running through their yard. Huge boulders of clay and ash were strewn throughout the landscape (Kennedy, December 24, 2008). James Schean wouldn't be as lucky. He awoke in his bed to the sound of snapping trees, a thunderous roar, and breaking glass. His whole house cracked, then crumbled, torn from its foundation and thrust nearly forty feet (Barker, January 9, 2009). As emergency response crews rushed to the scene, they saw a train heading for the coal plant come around the bend of the river and crash into the ash debris strewn across the tracks. That no one was killed was a miracle; that the crash did not occur during daytime commute, a stroke of luck in the midst of calamity.

Restricted Information

While a few residents found out about the spill from response crews, most residents discovered it as they drove to work. Three roads were closed, forcing many people to take long detours to their destination. Diana Anderson, who had dreamed of roaring trains and helicopters, realized something was amiss when she awoke. She grabbed her camera as she left her house for work. Officials along the roadside told her that there had been a small *mudslide*. TVA issued a brief statement downplaying the incident and urging residents not to drink tap water for a couple of days. Diana knew better and called local news outlets to inform them that it was not a mudslide but an accident at the fossil fuel plant. They ignored her and continued to report TVA's version of the disaster. Neither she nor her neighbors received any direct communication from TVA. In fact, it was not until December 30, 2008, that the TVA finally held a public meeting to inform residents about the spill. Along with a small response team from the county, the environmental group United Mountain Defense was on the scene within fourteen hours distributing 500 gallons of bottled water. It would be days before representatives of TVA began visiting some neighborhoods, and they did not get to many for weeks.

TVA tried to minimize the disastrous event. In the media and on its website, the Authority referred to the disaster as an "ash slide," assuring the public that the water was safe and the ash harmless and inert. Early statements would underestimate the actual extent of the damage considerably. An estimated 1.8 million cubic yards of coal ash was reported spilled, but TVA was forced to issue a correction when radar analysis later revealed the amount to be 5.4 cubic million yards. The explanation was startling: "Folks wanted to know. We sent somebody to make an estimate. There was no science behind it," said Ron Hall, manager of the Kingston Fossil Fuel Plant. Angela Spurgeon, whose dock was damaged by the spill, told a New York Times reporter, "That's scary to know they can be off by that much. I don't think it was intentional, but it upsets me to know that a number was given of what the pond could hold, and now the number is more than double" (Dewan, December 27, 2008).

TVA press statements referred to the disaster as an "incident." In a letter to affected home owners, TVA CEO Tom Kilgore referred to it as an "inconvenience." On another occasion, TVA senior vice president for environmental policy, Anda Ray, "refused to call the spill an environmental disaster," since in her mind coal waste is *inert*. Instead, she described the disaster as, "a challenging event to restore the community back to normalcy" (Flessner and Sohn, December 27, 2008).

Only one day after the ash spill, Tom Kilgore prematurely described the situation as "safe." He stated that "chemicals in the ash are of concern, but the situation is probably safe," long before TVA had any scientific evidence to support the claim. Similar to the Authority's first assessment of the amount of coal ash released, no science was behind his claim—although that fact was not made clear in Kilgore's statement. The Authority's analysis of its samples was not complete until three weeks later. At the time, Kilgore reassured the public, the TVA had neither collected nor tested environmental samples. Nevertheless, he boldly declared, "we don't think there is any immediate danger." The TVA continued to deny categorically that either the public or the environment were in danger. Just a few days after Kilgore's statement, Gilbert Francis, Jr., a TVA spokesperson, told the press that the ash spill material, "does have some heavy metals within it, but it is not toxic or anything."

In the immediate aftermath, the TVA might have simply admitted it did not know whether a public health threat existed and assured the public that every effort was being taken to make that determination. Rather than err on the side of caution, however, the TVA apparently assumed the situation was benign. Instead of warning the public to be cautious, they simply instructed nearby residents to wash their hands and avoid the affected area. Just take a few simple precautions and everything should be all right was TVA's mantra.

A week passed before TVA, EPA, and the Tennessee Department of Health (TDH) issued a joint statement that recommended that the public should avoid direct contact with the ash. Furthermore, children and pets should be prevented from the affected areas.

Sandy Gupton, a registered nurse who hired an independent contractor to test her family's spring water, was quoted in the media: "They think the public is stupid and that they can't put two and two together" (Dewan, December 30, 2008). She and others were angry that a week went by before the TVA responded to their concerns. The criticism that TVA withheld information from the public would become a constant theme throughout the crisis. The TVA, however, was not the only agency that was under fire.

The timing of the release of the generic statement is curious, to say the least, in light of the events that preceded it. On December 28, the EPA issued a statement that water quality tests indicated elevated concentrations of heavy metals, most notably arsenic. The EPA noted that its samples did indicate heavy metals present in the water "slightly above drinking water standards," but below what is considered harmful to the public health (Dewan, December 30, 2008). That is, with one exception: arsenic was detected at levels considered "very high." The Agency would later release data showing the arsenic levels to be 149 times higher than the normal limit.

There was an enormous discrepancy between this statement and an earlier one issued by the TVA that stated that tests of the Emory River water showed elevated levels of lead and thallium. Spokesperson John Moulton said these elements could cause birth defects and reproductive system disorders, but posed no threat to the public because they are filtered out in treatment plants before being consumed by the public. In the same statement, he stated that the TVA found "barely detectable" levels of both mercury and arsenic. This statement puzzled and troubled residents since the TVA and EPA test results proved levels of these elements to be considerable.

The controversy continued. A few days after the spill, groups of dead fish were found in pockets along the riverbank. The TVA claimed that the fish had been stranded on land by the spill and had died; however, many people, including some fish biologists, speculated that the fish died from the high arsenic and mercury levels released by the spill. A spokesperson for the TVA stated that the arsenic levels were high because the EPA's measurements had soil in the samples. In the days and weeks that followed, as the TVA continued to play down the threat to the environment and to the public, other independent testers reported information to the contrary. For several days, TVA officials asserted that the fly ash sludge was not toxic in spite of the fact that, as noted in the *New York Times* and elsewhere, fly ash has long been recognized to contain dangerous levels of heavy metals. A 2007 TVA inventory filed with the EPA showed that in only one year the plant's fly ash byproducts included "45,000 pounds of arsenic, 49,000 pounds of lead, 1.4 million pounds of barium, 91,000 pounds of chromium, and 140,000 pounds of manganese" (Dewan, December 30, 2008). The fact that the fly ash impoundment in Kingston contained almost five decades of toxic byproducts added to the concern of nearby residents.

Risks Associated with Coal Ash

Coal ash is considered a hazardous substance, not a hazardous waste. However, in 2000, the EPA considered it the latter and wanted to issue guidelines for its disposal. Political pressure from the coal industry made the Agency back off. According to the Natural Resources Defense Council (NRDC), a 2007 EPA draft risk assessment indicated that some types of "coal ash disposal sites pose a cancer risk about 1,000 times the level considered acceptable by the agency" (Natural Resource Defense Council, 2009).

Penny Dotson, one of the affected residents, countered with the criticism that "hazardous" is "hazardous" whether you label coal ash a substance or waste. Many residents found this statement self-evident and to the point. Sarah McCoin, whose family has lived in the area for several generations, said, "If fly ash is not hazardous, then why are all these precautions being made?" (Mansfield, January 23, 2009).

Concern about the potential threat of the coal ash mounted as a study in *Scientific American*, titled "Coal Ash Is More Radioactive than Nuclear Waste," was cited widely in the media (Hvistendahl, 2007). Among other points, the article mentioned an earlier paper that appeared in *Science* that examined the uranium and thorium content from coal-fired plants in order to investigate just how harmful leaching could be. The study found, according to the *Scientific American* article, that "estimated radiation doses ingested by people living near coal plants were equal to or higher than doses for people living around nuclear facilities."

People's worries turned to the uncertainties surrounding drinking water, dust contamination, heavy metals, radioactive compounds, possible contamination from ash sludge on their property, the potential exposure of children and pets to toxics, and the effects on their own health as well as those of the several hundred cleanup workers. It wasn't long after the spill that residents close to the spill began complaining of health problems. Typical of these complaints is what one couple reported: "We have constant headaches, frequent nosebleeds and persistent coughing" (Roane County couple, February 10, 2009). During the entire interview, one or the other was plagued with coughing fits. Other people reported increased incidents in asthma attacks while others reported having asthma symptoms for the first time.

Dotson, a former nurse from Illinois, and her grandson Evyn lived in a trailer that had a magnificent view of the river. Evyn was born with cerebral palsy, and Penny's worries about her grandson's health increased after the spill. Three weeks later, as Evyn's health deteriorated, his physician advised Penny to move from their home immediately. The TVA relocated her first to a motel and then a house.

Environmental Testing Increases Uncertainty

Uncertainties and doubts increased as independent researchers began reporting test results that conflicted with the TVA's assurances. United Mountain Defense (UMD), a nonprofit environmental group concerned with mountaintop removal, located in nearby Knoxville, Tennessee, collected samples from the site and requested that scientists at Duke University conduct tests on the soil and water. The tests revealed that the ash spill did contain radium and arsenic at levels that Duke scientists argued could be deleterious to human health. In response, a scientist at Oak Ridge National Lab countered that the levels were not high enough to pose a risk.

The Duke University team would, later in May, publish online in a peer-reviewed study that appeared in the *Journal of Environmental Science and Technology* that was conducted in a follow-up to the earlier study in conjunction with medical researchers from Duke and environmental engineers from Georgia Tech. The study's analysis of the ash samples showed that the sludge contained high levels of toxic metals and radioactivity. The study underscored other studies and observations that confirmed that the wet sludge was less of a risk to humans, but that when the sludge dries there is an increased risk from the possible inhalation of the fine particulates if they become air-borne. Thus, there could be a severe health impact on the surrounding population or the clean-up workers (Lucas, 2009).

Additional conflicting reports surfaced in the beginning of January when another environmental group coordinated research with scientists from Appalachian State University. The preliminary tests were said to show concentrations of eight toxic chemicals that ranged from twice to 300 hundred times higher than what is considered safe drinking water limits. The water samples, conducted according to EPA specifications, were said to have elevated levels of arsenic, barium, cadmium, chromium, lead, mercury, nickel, and thallium. Cadmium levels were detected two and a half times the allowable level for drinking water and four to seven times higher than the maximum level for wildlife. Lead was found to be two to twenty-one times higher than allowable limits for drinking water and nearly sixty times the maximum level for aquatic biota (see Smith).

In a press release issued by Appalachian Voices, Robert F. Kennedy Jr., Chair of the Waterkeeper Alliance, was quoted as saying: "Although these test results are preliminary, we want to release them because of the public health concern and because we believe the TVA and EPA aren't being candid" (Appalachian Voices, January 1, 2009).

Dr. Shea Tuberty, an environmental toxicologist at Appalachian State, who was one of the lead researchers, made an alarming statement in the same press release. Tuberty argued that both the TVA and EPA know what is in the ash by virtue of their routine solid waste disposal procedures and should be familiar enough with the numbers specifically concerning heavy metals that were released in the spill. Tuberty went on to say that there was sufficient information for them to calculate the numbers of tons of heavy metals that were released in the water and concluded by saying, "I think it is going to be a frightening number" (Appalachian Voices, January 1, 2009).

As the discrepancies between TVA's test results and test results reported by independent researchers and the EPA mounted, increasing criticism of TVA's motives and methods was made. The variability in results that was being reported by third party sampling became especially troubling to both environmentalists and some local residents. Soon accusations were made that TVA might be skewing the location of the water sample sites. The Nation's Institute Investigative Fund showed independent experts the GPS locations of the TVA's testing sites. *Nation* magazine reported the experts' response to the data. Hydrologist Bob Gadinski, argued that TVA's reliance on only five test sites was too limited. Moreover, he made the criticism that the testing at single depths rather than multiple depths at each test site was inadequate.

Representatives from the Environmental Integrity Project (EIP), United Mountain Defense, and a biology professor from Appalachian

State University contended that the site locations were "intentionally biased for nonsignificance." Gadinski asserted that at least three of the five sites chosen were points in the Emory River where ash contamination was less likely to reach. Paul Stant of EIP was quoted as saying, "They either didn't know that this testing scheme was poorly designed, or they knew and didn't care" (Hearn, April 2, 2009). While it is impossible to know what "they" knew or did not know or what their motivations were by this time, many local residents seriously doubted TVA's veracity. While there was much concern about the possible contamination of the river and of drinking water, the greatest threat to public health was dried coal fly ash becoming airborne. While there was little threat of that happening to the ash in the Emory River, there was considerable concern about the fly ash becoming airborne on land, especially with Eastern Tennessee's notoriously hot and dry summers. The TVA dismissed the fear of what could happen once the winter rains stopped because the Authority was confident that the cleanup would be completed before that time. TVA had no idea what it would be up against: It had seriously underestimated how long it would take to complete the cleanup.

The Cleanup Effort

The TVA assured the public the ash would be disposed of in just a few months. However, it was then faced with the unexpected reality that it might be more like a few years. In the meantime, helicopters with large buckets—similar to those used to fight forest fires—flew over the disaster site and dropped eighty-five million tons of grass seed. Over the seed, they sprayed tons of a dust suppressant. They then dropped tons of straw over the area. This activity resembled a war zone with helicopters flying overhead day after day. As many locals predicted, this approach had little effect. The unusually cold winter prevented the grass from sprouting, and the dust suppressant dried and peeled away into thin plastic strips. Despite the seemingly Herculean efforts, dust was everywhere: over the roads, on people's homes and property, and on cars and pickups, which became the main conduit by which the dust was spread near and far.

As the disparity in the risk evaluations grew and residents learned more of the potential health risks, concerns about health increased. Some families living close to the disaster stated that they were "shocked at the news of the spill. Now they are terrified," given the early EPA tests results showing the presence of elevated levels of arsenic in the water (Keil, January 2, 2009). "If you don't know the answer, don't lie to us. Tell us you don't know the answer, it is a lot easier to take," proclaimed a grandmother whose grandchild's health was affected (Keil, January 2, 2009). Another resident expressed the fears of many when he said, "You can't see what heavy metals might be doing to you. It is real scary not being able to see and know what is going on" (Keil, January 2, 2009).

Conflicting reports mounted and rumors spread; some old timers told the story that when TVA was cleaning out asbestos from some of the old fossil fuel plants, they dumped the waste into the ash sludge heaps. In addition to their concerns about the devaluation of their property, their inability to sell property outright, the potential impact on the environment, and the length of time it might take to clean up a spill of such magnitude, it was not long before many residents said that their number one concern was for their own health and the health of their loved ones.

As part of the cleanup effort, large trucks drove between the spill site and the quarry, carrying crushed rock to build temporary causeways on the spill site to allow heavy equipment to access the site. Many residents living in the Swan Pond road area, which was part of the spill site and near a granite quarry, complained that these trucks were strewing coal ash along the road and onto their property. TVA, first denying and then admitting the problem, ordered truck washing machines to wash the truck before they left the site. Several weeks later, the truck washer arrived. Apparently, TVA had erroneously ordered warm-weather washers, not washers meant to function in an unusually cold east Tennessee winter. The truck washing precautions were delayed, causing increased concern from local residents. Once the truck washers arrived, concern continued. A number of residents witnessed, and in at least one case videotaped, trucks leaving the site without being washed.

Safety of Cleanup Workers

An added concern for many was the health and safety of the cleanup workers who were working 24/7. They had not been issued the proper clothing and respirators. Many people witnessed the crews working on site with few, if any, precautions. At one point, at least one worker made a complaint about potential worker exposure to high-level background radiation. The complaint prompted the TVA to hire a consultant to identify hot spots and initiate proper precautionary measures. Health and safety concerns for workers continued until at least June when, in a community meeting with TVA, the EPA, and TDEC, a women rose from her seat and approached the microphone to express concerns for her husband's and her family's health because he had not been provided the proper protective clothing and equipment. She proceeded to describe a classic concern among families who work at toxic sites. Because proper safety clothing was not provided, her husband had to wear home his ash-laden boots and clothes, which she washed in the family washing machine, thus patently exposing her children to fly ash. Seven months after the spill, a spokesperson said that TVA had begun working on the problem and it would be remedied. Many people in the audience found the TVA's acknowledgement that a simple worker safety problem had been ignored so long incredulous.

Downplaying Disaster and Corporate Credibility

TVA's credibility and response to the spill was further eroded when an internal TVA memo prepared by the Agency's public relations staff, labeled, "risk assessment talking points," was *inadvertently* emailed to the Associated Press (Mansfield, January 23, 2009). The memo stated that the coal ash spill was best described as a "sudden accidental release," rather than a "catastrophe." The memo also was revised to remove "risk to public health and risk to the environment" as the reason for monitoring water quality. References to "toxic metals" were moved to a section dealing with water monitoring, and a discussion of fly ash was revised to state that it consists of inert materials not harmful to the environment.

TVA spokesman, Tom Moulton, in response to the apparent attempt to downplay the disaster, said, "We were putting in the word we thought … best described it at the time. Now, we certainly realize it was a very serious event and we realized then it was a serious event. There was no attempt to downplay it" (Mansfield, January 23, 2009). The AP story also underscored how even local officials were beginning to have serious doubts about TVA's credibility. A spokesperson for Roane County Emergency Agency, the county in which the disaster occurred, stated that his agency contacted the Tennessee Department of Environment and Conservation (TDEC) and the EPA as soon as they could for a second opinion on the severity of the disaster: "[W]hatever TVA was saying, we didn't feel like the public had confidence in what their reports were saying" (Mansfield, January 23, 2009). Not only were many people growing increasingly concerned about the Authority's lack of forthrightness, many members of the press were as well.

One journalist in an off-the-record aside to this author stated, "As much as I would like to take their [TVA's] word at face value it is apparent that they are purposely withholding some information and not leveling with the press" (Unnamed Journalist, January 12, 2009). In fact, a later TVA internal report would state at least in one documented incident that "repeated efforts by the media to learn anything about TVA's culpability were met with artful dodges" (Tennessee Valley Authority, July 23, 2009). One affected resident early on made a similar observation: "TVA tends to dance around the issue and not tell you direct answers" (Roane County Resident, February, 2009). Another resident, who had attended several public hearings, stated, "TVA will sand bag you with tons of irrelevant data but not answer your question" (Swan Pond Road Resident, February, 2009). Yet another resident, whose property was covered with toxic sludge, complained that it took more than a month for TVA to release material data sheets on the ash, alleging that the delay was "purposeful." Another person asserted, "TVA hasn't been honest with people. They have told so many different stories and they have not been consistent. It is pretty hard to believe anymore anything that they say" (Kingston Resident, February, 2009). One mother of small children sighed, "I don't know who to trust anymore" (Roane County Mother, February, 2009).

One blogger, whose home was damaged by the spill, expressed his distrust in TVA: "And at this point in time, your credibility is already shot due to your spokes peoples' previous misleading statements and half-truths on timelines, leaks, dead fish and toxicity, not to mention their pontificating on the cause of this completely avoidable disaster" (mysuburblife.com, January 15, 2009).

Many journalists complained that it was difficult to gain access to some TVA officials. Some reporters said they had great difficulty reaching the Authority via telephone. Others complained that usually they were only allowed to talk with TVA public relations officials.

Trust in the Authority was further diminished when it was revealed in July 2009 that the field notes of the inspector who last inspected the failed fly ash pond stated that while he was away from his Chattanooga office responding to the spill, his notes disappeared from his desk. Chris Buttram, who had never inspected a fly ash pond, stated that after inspecting the pond he returned to his office and left them on top of his desk where, according to him, they remained until December 22, 2008. Almost equally shocking is that in his deposition, Buttram not only said he had never conducted such an inspection before, he also claimed that he was not provided with a criteria for inspecting the pond. Furthermore, he could not recall having seen any ash pond design drawings before he conducted the inspection.

Several months later, the doubts and suspicions held by the public and the media were confirmed by the TVA's inspector general's (IG) report that stated that not only had the TVA ignored several decades of warnings that could have prevented the tragedy from occurring, but that it had also made a conscious decision to suppress facts. In commenting on the TVA's failure to investigate and report management practices that contributed to the tragedy, the IG's report states:

> [I]in fact, the TVA would not review what management practices may have contributed to the failure, but would instead tightly circumscribe the scope of review to intentionally avoid revealing any evidence that would suggest culpability on the part of TVA. In fact, it appears that TVA management made a conscious decision to present to the public only facts that supported an absence of liability for TVA for the Kingston spill.

TVA's Dilemma: Accountability or Litigation Strategy? TVA had a clear but difficult choice to make in the aftermath of the Kingston Spill. One choice was to conduct a diligent review of TVA management practices as well as to conduct a technical physical examination of the failed structure and then to publish whatever was discovered to the world. The second choice was to "circle the wagons," carefully craft press releases to project TVA in the most favorable light, and to tightly control any reports done by TVA of the failure to minimize legal liability. The first choice required a value judgment that a government agency causing a major disaster affecting the lives and property of citizens around the Kingston Fossil Fuel Plant should err on the side of transparency and accountability. The downside to this choice is providing fodder for plaintiffs in litigation against TVA and bringing perhaps additional scrutiny on the agency.

The second choice also required a value judgment. That choice placed a premium on the preservation of TVA assets and the protection of an image of environmental stewardship. The advantage of this choice was limiting legal liability which arguably inures to the benefit of ratepayers and avoiding scrutiny of TVA management practices that might have contributed to the Kingston Spill.

We are not privy to the calculation made by TVA as to the relative merits of these two difficult choices. We are, however, privy to facts that suggest a predictable outcome from TVA electing to go with the second choice. (Tennessee Valley Authority, July 23, 2009)

Criticism was made and many questions were asked about the TVA's response to the disaster: Why had the TVA not immediately informed nearby residents of the disaster? Why were the people who lived close to the spill site or on whose property ash sludge had settled—and there were many—not evacuated as a precautionary measure? Why, in fact, had the TVA immediately assured people that they were not in harm's way before any preliminary tests were conducted? TVA was also severely criticized for not having a disaster response plan in place to deal with an event of this nature.

The management culture of the Authority came under increased scrutiny, and a later internal document would criticize TVA culture in its approach to ash management. The report found that there had long been an attitude in TVA management "that ash was unimportant." It went on to identify "significant weaknesses" in the Authority's ash management practice including, "failure to implement recommended corrective actions that could have possibly prevented" the spill, "inadequate communication practices," and the "failure to follow engineering best practices." Perhaps most telling are comments made by some current and former TVA employees. One manager stated, "Ponds have always been the back end of the plant." Another employee stated, in effect, that the ash ponds were located too far from the plant to be of much concern to management, especially since they were not directly related to power production (Tennessee Valley Authority, July 23, 2009). TVA came under further criticism for its initial response to the disaster. It was revealed that it did not have an incident response team in place to respond to the disaster that would have allowed it to work effectively with the regional and national responders.

Criticisms alleged that the TVA's response was not well coordinated. It had difficultly interacting with the preexisting structures established under the National Response Plan established by Homeland Security. TVA offered the excuse that because it was technically a public corporation and not a government agency, it was exempted from having to conform to the plan. Nevertheless, it was difficult at first for county, state, and federal responders to interact officially with TVA because of its nonconforming response structure. Some individuals were critical of TVA for lacking any kind of well-coordinated response team. TVA remedied this situation a week later by hiring a private consultant to establish a plan consistent with the national response plan. Some critics saw this initial response and delay as indicative of a bungled response.

TVA was not the only party blamed for a tardy response. The Tennessee Department of Health also came under the scrutiny of some public criticism for its perceived failure to respond. As concerns about public health increased in the initial days and weeks following the disaster, a number of affected residents became increasingly preoccupied with health concerns. Many felt that there was a small window for the medical screening for the exposure to heavy metals and arsenic and demanded either the TVA or the TDH initiate and pay for screenings of hair, blood, and urine. Concerned citizens wanted to establish an early baseline by which to judge further exposure to potentially harmful substances. The costs of the tests, around $500 to $700, were prohibitive for most residents. Urgent emails and public flyers were issued to appeal for donations. Eventually, approximately twenty individuals were tested and some tested positive for heavy metals or arsenic. TDH's response to the call for biomonitoring was dismissive: "The existing data show that there has not been a significant exposure pathway to humans breathing air, drinking water and the water treatment plants or private drinking wells. Most people do not have extended physical contact with the ash. Therefore TDH does not recommend biomonitoring in the community near the fly ash release" (Tennessee Department of Health, February 13, 2009).

This statement was surprising because it was released in early February 2009 when test results were still being reported by independent researchers that contradicted TVA's testing. Given the disparity in tests results, it was also premature. At this point, testing was inconclusive, at best. Since the TDH had conducted no biomonitoring of its own, how could it be certain that there was no potential harm? How could it assure the public that no public health threat existed? Neither the TVA nor TDH had any ironclad information to support this assertion. In the same "coal ash release fact sheet," "emergency" and "disaster" are not mentioned. A "release" sounds more like a natural event than a man-made disaster—TDH based reassurances, in part, on soil and ash testing. It collected twenty-nine soil and ash samples. Sixteen soil samples were from impacted residences as well as some background samples outside the impacted area. It is unclear if the total number of sixteen samples is impacted properties or also includes properties outside the impacted area. The remaining thirteen were samples of ash. Considering the size of the affected area and the number of properties contaminated by the spill, the total number of samples is small, indeed (Tennessee Department of Health, February 13, 2009).

TDH played down the potential harm to public health and reassured the public long before enough empirical evidence was available to warrant such assurance, a strange position for a state agency whose primary mission is to ensure public health. To many people, TDH's response looked more like a rush to judgment than a scientific assessment.

TDH came under criticism for its failure to participate, but mostly because its response seemed delayed to many residents. TDH and the Agency for Toxic Substances and Disease Registry (ATSDR) claimed that they were on the scene within twenty-four hours inspecting the site and later reviewing environmental data test sampling. However, they themselves did not conduct a thorough investigation of nearby residents for several weeks.

Approximately three weeks after the disaster, TDH's staff interviewed 368 residents that lived within a mile and a half of the spill. The health survey did little to ameliorate the concerns of most residents. A household survey, not a physical examination or testing, was conducted, despite the fact that 33 percent of those surveyed said they had since the spill experienced coughs, headaches, wheezing, and shortness of breath. Approximately half of the respondents reported mental health issues such as anxiety and stress. The TDH responded to these complaints with a press release downplaying the effects of the spill and stating the majority of the respondents reported no change in their health status (Tennessee Department of Health, February 13, 2009).

While this characterization of their findings is true, it ignores the actual number of people who otherwise reported. Too many who lived in the area felt that the report ignored their concerns. In many ways, as the TDH admitted, the report was inconclusive. Some felt the report was conducted too late and was too little. Others reported that they expected a more proactive health assessment that would include physical screening rather than a survey about how people felt. Some, who lived outside the mile-and-a-half test area, felt that they had been neglected and that their health concerns were as legitimate as those interviewed. In early June 2009, at a public meeting that was held with EPA, TVA, and TDEC officials, one woman expressed a similar sentiment: "They say we are not affected in our zone, but if you look they have that circle, if that highlighter had been a sixteenth of an inch larger than it would have been on top of our house. ... TVA says I am not affected, I have never been so affected by anything in my life" (WTVC News Channel 9, June 24, 2009).

TVA's credibility and increased uncertainty continued in mid-September 2009, when residents adjacent to the Kingston Fossil Fuel Plant were showered with a coal ash snowstorm that covered their cars, homes, and property. Randy Ellis, a Swan Pond Road resident, first called the TVA to inquire about the ash. He was told to contact the TDH. When he was told that the lead TDH contact person was out of town, he called the Air Quality Control Board in Knoxville. He also called Mike Farmer, the Roane County executive, and Chris Mason, the Harriman City mayor, who happened to be in the vicinity. As the two men arrived, the ash started to fall again, so much so, that Mason had to wipe it from his head.

Eventually a TDH team from Knoxville arrived and collected samples, but told the men that they had no idea what was going on. Less than an hour later, the TDH crew returned with Leslie Nale, the plant manager. Nale informed the men that TVA had acquired a permit from the TDH office in Nashville to burn a higher sulfur coal in the present stacks in order compare the effectiveness of the old stacks with the new scrubbers that were to be installed. Randy Ellis said he later called Nale to ask if anyone in the community had been informed of what was going on. She replied, "No." Ellis then asked her what the health effects of such an incident might be, and she replied that she didn't know. Ellis later stated: "WOW, that really made me feel comfortable!" (Marcum, October 2, 2009). When TDH notified TVA of the incident, the unit burning the higher sulfur coal was shut down. Both the TVA and TDEC stated that they were unclear how the fly ash was released (Marcum, October 2, 2009). Steve Scarborough's statement resonated with many voices in the community, "The current TVA plant manager ordered the tests shut down as soon as we notified them, but they never spoke to the obvious question of why we had to notify them their test was screwing up and snowing coal ash on the community" (Roane Views, October 2, 2009).

Concerned about the potential short- and long-term health impact of the ash spill, many community members (including the author) and nonprofit organizations requested the ATSDR conduct a health assessment. When released, the health assessment raised serious questions about the integrity of the report and the Tennessee Valley Authority's role in writing the report.

The report raised several basic questions. For instance, rather than quelling health concerns over the TVA ash spill disaster, the report perpetuates long-standing concerns about the TVA's credibility and its role in the clean-up efforts as well as the appearance of a questionable relationship between the TVA and TDH. More important, beyond these basic questions loom larger ones that question how defensible an environmental science is without a transparent review process. For that matter, how dependable is a government-controlled science that collaborates with environmental polluters and avoids directly confronting potential harm while dismissing a community's concern and downplaying threats to human health? In short, how valid is a science that reassures the public that little or no harm has been done while serving to protect the interests of a select few?

Skeptics now wonder if TVA's past tendency to adopt a litigious strategy over accountability may have played a role in the issuance of the TDH health assessment report, which basically plays down any present or potential health threats. The TVA not only supplied much of its own data, on which the report was based, but actually was given the *courtesy* of reviewing the draft of the report and making changes that TDH depicts as *minor*. After TVA's review comments and changes were made, other agencies were also allowed to comment. In an interview with the *Roane County News*, the TDH spokesperson could not say what specific changes were made by the TVA.

The health assessment report raises other serious questions as well. First, the report mirrors statements made by TDH in the early hours following the disaster (prior to the conduction of any substantial testing) by stating, in effect, that TDH doubts any adverse health effects occurred. TDH's report also seems to imply that there is little long-term health threat, even though it can often take years to detect evidence of the deleterious effects of such toxins and longer to document the effects. While long-term health effects cannot be predicted with certainty, to dismiss the possibility of such deleterious effects is, nevertheless, premature.

One of the most astonishing points contained in the report states, "The Tennesssee Department of Health concludes that the data collected by non-governmental organizations was of limited usefulness in establishing the long-term health implications of the coal ash release" (Tennessee Department of Health, December 9, 2009: xxxi). However, ironically, the report continues by stating the "non-governmental data confirmed data collected by governmental agencies."

First, it needs to be noted that the TVA, in the early weeks after the tragedy, did everything within its power to prevent independent researchers from gaining access to the site. Despite this deterrence, some organizations did have limited success in obtaining and analyzing test samples. Although the report somewhat coyly states that some of this data confirmed governmental data, it seems to dismiss the fact that such independent testing also raised valid concerns about the health and environmental impacts that were at variance with what the TVA and some other agencies were reporting. This apparent dismissal of nongovernmental research raises the question of what criteria were used by TDH to dismiss the usefulness of independent testing. Furthermore, was this same criteria also used to evaluate the tests conducted by the TVA? If the TDH admitted even limited value of the nongovernmental research, why were those organizations not given the same *courtesy* afforded the TVA to preview and amend the report? In other words, why were independent researchers who had conducted research on the spill, who have no ties to the EPA, not allowed to review and comment on the report prior to its release for public comment?

The report also glaringly ignored findings released by the Environmental Integrity Project (EIP) highlighting a much grimmer picture of the amount of toxins released in the spill. Among other assertions, the EIP reports that an analysis of the TVA's own data shows that a total of 2.66 million tons of ten toxic pollutants—arsenic, barium, chromium, copper, lead, manganese, mercury, nickel, vanadium, and zinc—were released by the spill when an estimated one billion gallons of coal ash were released into the environment. EIP compares this release with the discharge of only 2.04 million pounds of such toxins from all U.S. power plants into surface waters in 2007. These findings are troubling because many of these metals bio-accumulate and can pose significant risks to human health. This concern raises a serious red flag about the potential adverse effects of the TVA disaster. At least in the minds of some skeptics, it questions the TVA's continued downplay of potential long-term health and environmental effects.

Of equal importance is the fact that in 2007, EPA published a draft risk assessment, which found extremely high risks to human health from the disposal of coal ash in waste ponds and landfills—let alone uncontrolled catastrophic releases into the environment. The failure of the TDH review to consider these findings also underscores that the report ignores any discussion of the potential short- and long-term synergistic effects of being exposed to all, or some combination, of these toxins found in the coal ash. In the past, synergistic effects have been too long ignored in environmental health research. The TVA spill provides the scientific community with an ideal opportunity to explore these potential effects and correct this oversight.

It is also curious that the TDH acknowledges, as it did in an earlier report, the stressful mental health effects of the disaster on residents. Apparently, however, the TDH made no attempt to probe this finding. Thus, while the report recognizes the role of stress, it does not seem to investigate its potential impact in any significant manner. This shortcoming is profoundly disturbing: The mental health effects of disasters are well documented and recognized as playing a significant role in the lives of disaster victims. This oversight is also ironic—the Agency for Toxic Substance and Disease Registry (ATSDR) has in the past devoted considerable attention to the psychosocial impact of environmental disasters.

It is essential to remember that the original impetus for this health assessment was the request of the public and many NGOS to have the ATSDR, a federal agency, not TDH, conduct the assessment. However, as has so often been the case, the ATSDR, while involved in the report, handed it off to the TDH to conduct. This habit is noteworthy. Testimony by independent scientists and a U.S. Congressional report have recently criticized the ATSDR for delegating the role of such investigations to third parties rather than employing the independence and expertise of the agency. The Congressional report reminds the ATSDR that the agency's mission "is to serve the public by using the best science, taking responsive health actions, and providing trusted health information to prevent harmful exposures and disease related to toxic substances" (U.S. Congress, House, March 10, 2009). Once again, the Agency seems to be avoiding its mission by commissioning a state public health agency to conduct investigations, in this case of great national importance, which the ATSDR is better equipped to investigate and is mandated by Congress to conduct.

The same report also observes "across the nation community groups often believe that ATSDR has failed to protect them from toxic exposures and independent scientists are often aghast at the lack of scientific rigor in its health consultations and assessments." Some eastern Tennessee residents (especially those whose lives were most affected by the spill) are beginning to voice similar complaints.

Especially notable among its findings, the House committee report argues that ATSDR repeatedly seems to "avoid clearly and directly confronting the most obvious toxic culprits that harm the health of local communities." Instead, the report contends, the agency seems to trivialize and ignore legitimate health concerns of local communities and of well-respected scientists and medical professionals. One ATSDR scientist even told the House committee, "It seems like the goal is to disprove the communities' concerns rather than actually try to prove exposures." Almost one year after the Congressional hearings, similar criticisms are now being raised with the release of the draft of the TDH report.

As noted social scientist Lee Clarke has observed, researchers have long recognized that the determination of acceptable risks is usually a political and not a purely scientific process. Risk assessments are made within organizations and agencies, which attempt, in Clarke's words, "to shape and form what constitutes an acceptable risk, and bargain among themselves as to what is acceptable to them" (Clarke, 1989). As Clarke goes on to note, too often the result is not to the advantage of that portion of the public that bears the disproportionate burden of such risk assessments. Instead, as in this case, the very government agencies established to protect the public and the environment dismiss public concerns while protecting the interests of industrial polluters under the guise of *pure science*. The public, thereby, is left with the question of how it can obtain a science that serves its interests above and beyond the interest of corporate polluters. In the chapters that follow, we will explore this and other questions.

In the following four chapters we will turn our attention to the multiple ways in which uncertainty becomes politically inflected by examining mediated accounts of disasters, the contestation of knowledge, the production of uncertainty, and finally the sequestration of knowledge. In the process, we will see how uncertainty is shaped, distorted, manufactured, and massaged in ways that exacerbate uncertainty and undermine our regulatory system. The result, thereby, makes both our citizenry and environment vulnerable to the ravages of disaster and less able to recover in a timely fashion.

7. Mediated Disaster Narratives

Does all this kind of thing happen so often that nobody cares anymore? Don't those people know we have been through enough? We are scared to death. We still are. ... Are they telling us it was insignificant, it was piddling? ... Even if there hasn't been a great loss of life, don't we deserve some kind of attention for our suffering, our human worry, our terror? Isn't fear news?

Don Delillo, *White Noise*, 1984: 161–62

If the media spoke with Exxon people first, and then came to us, we would find ourselves having to disprove what Exxon had already said, instead of just being able to tell our own story. And the stories were always spun toward Exxon. So I think it's a game of media capture, I think all these big corporations have the public relations angle completely down, a tool that they use to control damage and ultimately to minimize liability in court.

Riki Ott, marine biologist (Chelsea Green, March 24, 2009)

In a time of calamity, community members and their families struggle not only to regain control of their lives, but also to refute the objective frames of disasters that are offered by corporations, government agencies, and the media (Button, 2002: 144). News becomes the contested terrain in which a struggle for meaning is often encountered. The way in which the media frames an event strongly mediates how the public thinks about the event—including policy makers, politicians, and other influential parties that play a significant role in shaping and defining our societal responses to disaster. While some may argue that the media may not tell people what to think, the media does have the power to tell people what to think about (Wallack and Dorfman, 1993). The media also literally have the ability to recognize a disaster and legitimize it as an actual event or overlook or dismiss it as a nonevent. Or, as in the case of the Shetland Islands oil spill, if the anticipated media framing of the event fails, the media can abandon the story and go home.

For every disaster story like Love Canal, Three Mile Island, or the TVA ash spill, hundreds of similar events never gain media attention and come under public scrutiny. While the magnitude and severity of the event may, to some degree, determine such oversight, many significant, neglected disaster stories are never told.

The media tend almost exclusively to focus on melodramatic events, which serves to galvanize the public's perception that disasters are random, exceptional events, not as common and ubiquitous as they are in everyday life. Even those disasters that eventually gain national attention, such as Love Canal or the toxic groundwater contamination in Woburn, Massachusetts, can sometimes take years before they gain national prominence.

In the case of the Woburn story, although it was covered for some years by the *Boston Globe* and the local press, it did not come to the general public's attention until Jonathan Harr's bestselling book, *A Civil Action* (1995), won the National Book Critics Circle award for nonfiction and was made into a movie by Disney's Touchstone Pictures (1998). It took a highly sensational book such as *A Civil Action*, which was written in large part as a courtroom/suspense drama to gain national attention (Button, 2002). A decade before the publication of the book, journalist Paula DiPerna wrote a thorough account of the disaster as it unfolded. Despite the book's merits, it was largely ignored, perhaps because it was written as a case study. In 1990, sociologists Phil Brown and Edwin J. Mikkelsen wrote *No Safe Place*, a scholarly, but easily accessible account of the Woburn story that also gained little attention outside of academia.

Despite a best-selling book and movie about the groundwater contamination at Woburn, in many ways, Love Canal is a better-known disaster and considered to be a major benchmark in our nation's environmental history. While it could be said that Love Canal gave birth to the environmental movement, at least in the form of "not in my backyard" (NIMBY), it did not set the legal and public benchmarks that the groundwater pollution in Woburn established. Yet, the very fact that one has to qualify the Woburn story with the statement about groundwater contamination serves to demonstrate how relatively unknown it is compared to Love Canal.

What makes this comparison even more interesting is that as former EPA official Hugh Kaufman, who contributed to exposing the controversy surrounding Love Canal, has so trenchantly observed:

> Love Canal was a catastrophe, but it wasn't the worst catastrophe in the country, or even in New York State, or even in Niagara County ... it was Love Canal, Hooker Chemical, and Niagara Falls, honeymoon capital of the world. Editors all over the country had orgasms. So Love Canal became catastrophic, not any of the other 10,000 hazardous dumps across America. (Hoffman, 1997 citing Keating and Russell, 1992: 33; also cited in Fortun, 2001: 70)

Fortun, in her groundbreaking analysis of Bhopal Chemical, argues rightfully that Kaufman's analysis of disasters lacks an adequate explanation as to why so many of the 10,000 events that she alluded to simply remain incidents and did not become recognized events. She cites scholar Lee Wilkens's observation that the media's standard mode of analysis is to relegate these events to the local level, leaving out, in Fortun's words, "significant background information that would indicate translocal relevance" (Fortun, 2001: 79).

No doubt much truth is in this statement. As I have argued elsewhere (Button, 2002), the media's analysis is episodic, not systemic, and it, therefore, leaves out the kind of analysis that would place disasters in the larger context of society, translocal and otherwise. However, what this neglect also achieves is that it presents disasters simply as isolated events rather than part of common everyday life. The 10,000 other toxic sites and all other so-called incidents waiting to be recognized as events serve to make this abundantly clear. For the media to recognize this fact would require that they reveal the true social arrangements of our culture and the asymmetrical power relations that produce disasters. Such an approach would require exposing the underlying cultural logic that reinforces the hegemonic forces in our society.

Disasters as Isolated Events

In the case of the PBB contamination of cattle in Michigan (1973), which contaminated human breast milk, federal and state authorities played down the problem. The story appeared briefly in the *Washington Post* and the *New York Times* and then evaporated quickly from public discourse (Eggington, 1980). The PCB contamination of the Binghamton, New York, State Office Building in 1981 occurred shortly after the national and international attention that focused on Love Canal. Many of the same state health officials were involved in both disasters. It took almost fourteen years and over $53 million to decontaminate the building before it could be reopened. The event received some nominal national attention but never received publicity on the scale that surrounded Love Canal. Lee Clarke, who wrote a trenchant analysis of the event, suggested that local citizens were unsuccessful in gaining outside support or gaining attention from the media (Clarke, 1989: 155).

Another PCB story in New York State also received scant attention compared to other environmental disasters. In 1991, an off-campus traffic accident damaged electrical transformers in five buildings on the State University of New York at New Paltz campus. Five buildings on campus, including three dormitories and a new science building, were heavily contaminated with PCBs. Despite the fact that the accident occurred almost two decades ago, there is still controversy around the cleanup. The events surrounding the disaster received some modicum of national press but nowhere near as much as the event warranted.

When a disaster story does gain the attention of the national and/or international news media, the amount of time it remains a front page or top of the hour story is often severely limited even in those cases of catastrophic proportions. Stories are abandoned very quickly in the over-all scheme of things for other *breaking* stories. Even events like the Southeast Asia tsunami can rapidly dissolve from media attention. Despite the magnitude and severity of the event, eight months after Hurricane Katrina came ashore, the story was relegated to almost obscurity. Once such an elision occurs, it precludes any possibility, however remote, of a systemic analysis of long-term social or health problems. In the case of technological disasters, or disasters which have similar components, this treatment is unfortunate. The continuum of these particular phenomena often unfolds slowly and it can take years for many of the long-term effects to manifest themselves.

The catastrophic earthquake in Haiti (2010) gained considerable international attention for many weeks but then gradually receded from the front page, not before, however, an even larger earthquake hit Chile, which was 500 times larger in magnitude than the earlier Haitian quake. Despite the sheer magnitude of the quake in Chile, it did not, by any means, kill the estimated 200,000 people that the quake in Haiti did. Nor did it destroy Chile's infrastructure. Interestingly enough, the quake on the coast of Chile received comparatively little international attention before it, too, receded from the news.

To a limited degree, Hurricane Katrina may be an exception to this approach, although it can be argued that only certain media outlets—other than the local media that often persists in their coverage—such as the *New York Times* and National Public Radio (NPR) kept the story alive. Even in these cases, the media's focus is usually on selective areas and stories, such as the rebuilding of New Orleans. Vital topics, such as the status of thousands of evacuees who remain homeless across the nation, are often ignored.

How the Media Frames a Disaster

The way in which the media frames a story is of vital importance. The frames are the packages in which the central news story is developed and understood. Although, "largely unspoken and unacknowledged" (Gitlin, 1980: 7), they help journalists organize the world. They also strongly shape how we perceive the world. Frequently, these frames are based on unchallenged assumptions about the world. Rather than the cultural constructions that they are, frames appear to both reporters and news consumers as natural constructions. They present seamless, seemingly objective accounts of the given world, neither revealing their underlying subjectivity nor demonstrating the complex structure of disasters.

Steve Coll, an Associated Press reporter stationed in Haiti, summarized best the predictability of the news coverage of disaster in an interview with Bob Garfield, on the NPR radio show, "On The Media" (January 29, 2010). At the beginning of the interview, host Bob Garfield summarizes Coll's analysis:

> For instance, on day five you'll have your miracle story, someone pulled alive from the rubble, for example. On the first Sunday following the tragedy, there'll be the interpretation of meaning stories and lots of interviews with religious leaders. And on day 12 he calls

"heading for the exits," According to Coll, that's when editors begin to call their reporters home.

The contestation over the meaning of a disaster and the struggle to interpret it commonly result in a struggle to frame the event. The outcome is often dependent on who has both access to the media and a sophisticated understanding of how the media works. Unfortunately, corporations and government agencies have either the capital and/or the access that enables them to influence the media that disaster victims and their families do not possess. This influence is bolstered by the fact that the overall trend of most media reporting on disasters is the majority of voices that are published are interviews with such spokespeople instead of those disproportionately affected by the crisis.

One way to unpack frames of news-related events is to think of them as products of culture that have built-in assumptions about what categories are considered relevant. For instance, the "who," "what," "where," "when," and "why" of a news story are supposed to address what should be investigated—that is, the people defined as "who," "what" things are accepted facts, the geopolitical boundaries that constitute a "where," the time-line constituting "when," and the "why," or the explanation. The news media highly restrict this latter category as well as most others (Schudson, 1995: 14). For most news stories, the "why" is predicated solely on motives and thus precludes looking for the causes of disasters in broader social patterns.

News stories are, for the most part, anecdotal accounts and not systemic analyses of disasters and the larger historical circumstances that shape and determine them. In other words, such accounts ignore the disaster continuum on both ends and focus almost exclusively on the middle of the continuum, the triggering event, ignoring both the historical sequence of events leading up to the disaster and the long-term recovery process. This process collapses media accounts into a very narrow frame and ignores full explanatory approaches. It thereby reinforces the notion that disasters are exceptional events, which are not reflective of everyday life and our everyday material world that shapes them. This decontextualized scenario deters us from studying the nature of the social and cultural constructions of reality. The neglect of the longitudinal evolution of disaster serves also to reinforce the neglect of systemic forces, asymmetrical power relations, and the long-term impact of disasters on human communities.

The Framing of Disasters

If we view the construction of news from an arena broader than the newsroom and view it through the lens of Gramsci's notion of hegemony, the decisive role of the political/economic influence of certain interest groups emerges (Ryan, 1991). As one observer has argued, the "ultimate impact of framing is pro-establishment" (Iyengar, 1991). The underlying cultural logic of disaster narratives tends to reinforce the hegemonic forces of society. The discourse about disasters becomes, therefore, a discourse about the politics of disasters. Consequently, the extremely political process of controlling access to information and the construction of disaster narratives vary from incident to incident. Certain reoccurring themes emerge, reinforcing the hegemonic forces of society.

In so-called environmental disasters, the focus is generally on the effects on the environment and not the effects on human communities, however severe they may be. Many of the accounts of the *Exxon-Valdez* oil spill and, to a somewhat lesser degree, the accounts some twenty years later on the impact of the TVA ash spill are examples of this kind of framing that emphasizes a purely environmental perspective. This practice can be seen as an attempt to naturalize the disaster, even though, in both instances, the disasters were the result of human negligence. Many of the popular accounts of Hurricane Katrina resemble this pattern in that they ignore the man-made modifications to the environment, such as the channeling of the Mississippi River and the addition of oil and gas pipelines along the coast, that made the area vulnerable to disaster. This approach is taken a step further when disaster narratives place the events as ultimately being outside human control. Consequently, oppositional discourse about responsibility and blame of both government agencies and corporate entities is removed. The response to disaster is depicted as a valiant, and often futile, struggle with Mother Nature, creating a scenario beyond human control. The resultant reification makes the event appear more like an accident and less the result of human negligence (Tuchman, 1978).

Framing by Parties Involved in a Disaster

Very often those parties implicated in disasters are successful in framing narrative events that serve their interests while manipulating the facts of a story. One such case is an account told to me by the humorist Tom Bodett, who resided in Homer, Alaska, at the time of the *Exxon-Valdez* oil spill:

Well, I think the oil companies were getting their way with the media. They know how to use the media very well and how to feed the proper sound bites and things at the proper time, as well as how to stage a good production. For example, the outer coast [of the Kenai Peninsula] was pretty heavily oiled. Exxon hired sixty to eighty local people to go there and clean a beach. They brought in all these helicopters and put everybody in orange waist coats and flew them over there and put them on the skirmish line with their shovels. Dan Rather's crew came in from, I believe, CBS, and flew over there. They got some good footage over there for about a half an hour and they interviewed some very conscientious and concerned looking oil people and filmed them. Then they packed up their equipment and flew out. And that's what you saw on the evening news that night, Exxon's wonderfully orchestrated cleanup project. The next day those people were laid off and they [Exxon] never, ever, went back to the beach. It is that kind of stuff that is maddening to local people. (Bodett, interview with author, 1990)

The above incident was not the only time a beach cleanup was staged to influence the media. AP reporter Paul Jenkins tells of a similar scenario when Vice-President Dan Quayle, on his whirlwind tour of Prince William Sound, witnessed hundreds of workers cleaning a beach on Smith Island's north shore. A couple of days after Quayle's visit, Jenkins returned to the cleanup site and found the beach still covered with oil. The Exxon cleanup team had departed without notifying either the state or the U.S. Coast Guard. Curiously, according to Jenkins, the very day he witnessed the deserted beach, Exxon at a press conference in Valdez was claiming that the beach was still being cleaned (Davidson, 1990: 199).

Fact Checking and Past Disasters

The media can, however infrequently, be manipulated by people in affected communities that want to shape and embellish the story in such a way as to gain national media attention. Some environmental organizations in the wake of the TVA ash spill proclaimed to news organizations that the spill was larger than the *Exxon-Valdez* oil spill or framed the event as the biggest environmental/industrial disaster in our nation's history. Ironically, the *Exxon-Valdez* oil spill was not the largest oil spill in our nation's history. It is ranked as the thirty-fourth largest spill in the world. Prior to the BP Deepwater oil spill in 2010, the claim to the largest oil spill was given to the Brooklyn oil spill, which involved anywhere from seventeen million gallons (as compared to the official estimate of eleven million gallons in the *Exxon-Valdez* spill) to thirty million. Few people outside the New York City region have even heard of this spill (*New York Times*, June 18, 2007; *Mother Jones*, September/October 2007).

Surprisingly, many national media as well as local news organizations repeated the claim about the TVA spill without any apparent fact checking. The claim that the TVA spill is the worst environmental disaster in U.S. history is totally without merit, not only given the extreme magnitude and severity of the Exxon spill in relation to the TVA spill, but also given the fact that there are many other environmental/industrial disasters in our nation's history that far exceed the TVA spill simply on the basis of the loss of human life, which is a major measure of the severity and magnitude of any disaster. In a similar vein, some journalists have described the BP Deepwater spill as our nation's largest environmental disaster thereby ignoring the Dust Bowl of the 1930s or perhaps the Mississippi flood of 1927, let alone Hurricane Katrina in which some eighteen hundred people perished.

For instance, the 1907 mining disaster in Monogah, West Virginia, disaster claimed the lives of 362 miners. All the more ironic since it, like the TVA ash spill, was a coal mining related disaster. One of the worst industrial/environmental disasters is the long forgotten Hawk's Nest Tunnel disaster of the 1930s in which an estimated 700 men perished and hundreds suffered permanent respiratory damage from silica. Interestingly enough, according to Martin Cherniack's account (1986) of this terrible tragedy, it gained little or no public attention, aside from a New York literary journal, a radical labor press in Detroit, and a story written by an Associate Press stringer. The latter story, Cherniack claims, never reached farther than Charleston, West Virginia.

Arguably, the worst environmental/industrial disaster in U.S. history is the Libby, Montana, story involving W. R. Grace, the same corporation that was implicated in the groundwater contamination in Woburn. Local activists struggled for years to have their story told. The story first emerged on the national stage with reporter Andrew Schneider's groundbreaking coverage of asbestos poisoning in Libby (Schneider and McCumber, 2004; Peacock, 2003). Despite Schneider's award-winning coverage, two book-length journalistic accounts that followed, and an award-winning documentary, the story is perhaps not as well known as other disasters within the United States.

For over three decades, a vermiculate mine in this small Montana town supplied asbestos for most of the nation and much of the world. The asbestos claimed and continues to claim the lives of thousands of miners and their families as well as undoubtedly the lives of thousands of others around the nation and the world. While perhaps less known than Love Canal, Times Beach, or Woburn, the mining activities have resulted in not just another localized disaster story but in a national scandal, which took years of investigative journalism and community activism to uncover.

The story of asbestos mining in Libby is a classic example of the globalization of disaster where a disaster not only had lethal consequences for local citizens, but for people around the world. These victims came into contact with a host of asbestos products, from insulation to crayons. In an upcoming chapter, we will see in a startling way how the Libby, Montana, asbestos played a key role in one of our nation's worst tragedies.

These examples demonstrate how modern-day disasters can be distributed over space and time just as nuclear fallout from aboveground testing or radioactive contamination from Chernobyl. This nonlinear dimension makes it especially difficult for journalists to cover and for victims' voices to be located, let alone heard. However, these nonlinear dimensions allow social analysts to delineate and explore power as it is distributed through space and time.

Of course, what constitutes the worst environmental or industrial disaster is subject to varying definitions; however, despite such varying criteria, the TVA ash spill was without a doubt a terrible tragedy and of great significance given the controversy at present surrounding the coal industry. Nonetheless, it would not even qualify for the top ten of worst disasters under any of these definitions, and yet, the media unquestionably billed the disaster as the worst in our nation's history. Even major media outlets such as the *Nation, GQ Magazine,* and NPR mimicked the environmentalists' claims. When questioned about why they did not fact check the allegation, some journalists told me they just assumed it was true. When I described to reporters several different disasters which would tend to disprove such a claim, most responded that they had never even heard of the major contenders for the *worst* national disaster even though many had heard of the Exxon spill. Despite some familiarity with the Exxon disaster, most admitted that they never stopped to consider the comparison seriously. In fact, most journalists with whom I spoke admitted they knew little about the *Exxon-Valdez* spill or the circumstances surrounding it.

Media Resources

This failure illustrates, in part, the fact that reporters increasingly lack the resources and time to fact check stories even if they have the inclination. Equally important, few reporters have the scientific training to pursue an in-depth analysis of the controversies surrounding scientific uncertainties. Even if they did, such details are not perceived by their editors and producers as likely to sell stories and attract reader interest or, more important, generate media profits. The intricacies of such controversies do not lend the facts to an easily managed story. In many cases, they can evoke political controversies that the media want to avoid.

The failure of the U.S. government's untimely and inadequate response to Hurricanes Katrina and Rita did, perhaps, change both the nation's perception of the events and the news media's coverage. With the exception of *New Orleans Times-Picayune*, the *New York Times*, and NPR, an even less systemic analysis of the story might have occurred. The alternative media and internet blogs and coverage provided most of the in-depth, long-term coverage that these disasters warranted. However, how many people consume such coverage remains unknown. Nonetheless, the internet provided the voices of many victims and social commentators with a means by which to tell their stories and provide alternative narratives to the mainstream media.

As this is being written, the story of the earthquake in Haiti is unfolding. There are early signs that some of the lessons learned from these alternative narratives may have some effect on not only the news media's coverage, but also the long-term recovery effort to rebuild Haiti. On web media sites such as *Huffington Post, Common Dreams, Democracy NOW,* and *Counterpunch,* a systemic narrative is being provided. Signs of this influence may be seen in the Facebook initiated project, "No Shock Doctrine for Haiti," based on Naomi Klein's book, *Shock Doctrine* (2007). With other voices from various sectors, this project has gathered the strength of 150,000 individual voices to demand that Haiti be freed of its indebtedness by the international community. It has also demanded that Haiti be allowed to develop without the hindrance of oppressive neo-liberal strategies that would continue to make Haiti vulnerable to both chronic and acute disasters.

The Downsizing of Newsrooms

The ability of disaster victims and their communities to access the media and have their voices heard is now undermined, in the United States at least, by the significant changes occurring in American journalism. With the increased downsizing of the media and the loss of many local news outlets and the potential loss of major, national outlets like the *New York Times*, there is an increasing likelihood that more disasters stories will not be reported and those that are will not be covered in depth.

A perfect example of the potential long-term threat of our shrinking newsrooms is the laying off of Andrew Schneider from the *Seattle Post-Intelligencer*. The two-time Pulitzer Prize winning reporter, who was responsible for breaking the story of the decades-long asbestos poisoning of Libby, Montana, and the national scandal that surrounded it, is now an unemployed investigative reporter. He has his own blog but not the financial resources needed to cover the kind of in-depth reporting which made him famous. With scores of seasoned reporters like him in similar straits, how many Libby, Montanas will go unnoticed?

The increasing lack of public funding for public broadcasting is making matters worse because it is arguably, though not always, less propagandistic than commercial broadcasting. With the shrinking and disappearance of newsrooms and foreign bureaus, the likelihood of this fate is very real. With less access to independent information sources, disaster victims will have an increasingly difficult time getting their alternative narratives recorded and reported. As Nichols and McChesney argue, under these conditions, "Vast areas of public life and government activity will take place in the dark" (McChesney and Nichols, 2010). Whose voices will be heard and whose voices will be silenced will become an ever more pressing concern if more lights go out in pressrooms and studios.

Public Relations Fills the Gap

As newsrooms shrink under the economic pressure of the times, public relations has begun to fill the vacuum by manufacturing the terms of public discourse. Edward L. Bernays, considered to be the founder of the public relations profession, conceived of PR specialists as "hidden wire pullers" doing the work of the "intelligent few." Early in the twentieth century, Bernays espoused the "conscious and intelligent manipulation of the ... opinions of the masses," who he perceived as threatening to destabilize the ruling elite (Ewen, 1996: 167). Bernays held the notion that if public relations could understand the "group mind," then an interested party could "control and regiment the masses according to our will without their knowing it" (Ewen, 1996: 169). In those early days of spin, he actually proposed the fabrication of news. If Bernays were alive today, he might be surprised to learn how successfully his ideas have been applied.

Corporations responsible for the cause of disasters have long employed public relations tactics to improve their image and persuade the public of their version of events. Failing that, they generate enough uncertainty as to disarm critics. Exxon carried on a very aggressive, well-funded campaign effort in response to the *Exxon-Valdez* oil spill. Among the many tactics the oil giant used was to hire three British scientists to visit the spill and *independently* assess the damage. The scientists were chosen because they were considered friendly to the oil industry.

According to *Corporate Watch*, all three men were known skeptics of long-term ecological damage from oil spills. Moreover, they were reputed to believe oil spills do not significantly harm human populations. All three men had supposedly written on these subjects. Otto Harrison, Exxon's executive director of operations in Alaska, told an international petroleum conference that the experts were also chosen because Exxon believed that the public would find the men's statements more credible because of their English accents. In 1990, the consultants released a report stating the impact of the spill on the Alaska coastline was "likely to be short lived." In 1991, one of the men stated that as a result of the cleanup efforts and the scouring of winter storms, the coast of Alaska was largely free of oil by the spring of 1990.

As someone who traveled the affected coast of Alaska, I can attest that this was hardly the case. I saw large amounts of oil in several different places along the 1300 miles of affected coastline. Moreover, in 1992, NOAA estimated that in autumn of that year 12 percent of the total oil still remained in subtidal flats settlements and 3 percent on the beaches. Today, twenty years after the spill, oil is still present along the coast. The *Exxon-Valdez* Oil Spill Task Trustees expect direct damage to the wilderness to continue for decades. Scientists from the University of North Carolina, nearly two decades after the spill, estimate that it may take up to thirty years for the shoreline to recover.

The same Exxon consultants also stated at that time that the sea otter population was still abundant in the affected area and would rapidly reverse any losses sustained from the spill. This announcement was in contrast to three studies of sea otters in 1991 stating that chronic damages were limiting the recovery of the otters in Prince William Sound and that mortality patterns were abnormal compared to prespill data. By 1993, there were still no significant signs of recovery of the sea otter population.

Exxon undertook many other public relations efforts, including the production of *educational* videos such as "Scientists and the Alaska Oil Spill." Notice that the title itself attempts to play down Exxon's role in the spill. The video was offered at no charge to science teachers. The label on the tape says, "A Video for Students." The overall message of the video is that the spill was not as bad as had been portrayed. It attempts to convey the message that minimal damage was done to the coast and wildlife. The video also shows very little of the spill or its impact and, without any evidence to the contrary, downplays the spill. How successful the message was is probably unknown, but the video did make it into many school libraries and classrooms across the nation.

Another controversial video was produced to influence shareholders at the Exxon annual shareholders meeting. The thirteen-minute tape also downplayed the spill's effects and included scientists stating unequivocally that the affected area would recover. Its stated message was clear. "The brutal scenes of damage to Alaskan waters seen on nightly news programs were false."

Of course, Exxon was not the first oil corporation to undertake a massive PR campaign in the wake of a spill. After the Santa Barbara oil spill (1969), several oil companies, including Union, Mobil, Texaco, and Humble, launched a huge advertising campaign touting their environmental endeavors. Mobil ran an ad about the offshore blowout off Santa Barbara, playing down the incident and the impact on the environment: "Santa Barbara was a bad accident but not a disaster. ... The channel has long since been clean and so have the beaches" (Potter, 1973: 299). American Petroleum Institute, which had previously spent about $1.7 million on public relations annually, spent $9.4 million in the two years following the Santa Barbara spill (Easton, 1972: 219).

In 1969, the year of the Santa Barbara spill, Mobil Oil launched another media campaign to influence the public and Congress. In addition to its ambitious ad campaign, the company decided to bury its propaganda in prestigious contexts. They chose the op-ed page of none other than the *New York Times*, where, as one observer says, "they virtually owned the lower-right hand corner" (Sherrill, 1983: 60). This prominent placement gave them access unequaled by other oil corpora-tions. In a sense, they virtually had their own op-eds placed in the most well-read op-ed section in the nation.

Public relations campaigns on behalf of the oil industry in the wake of the BP Deepwater blowout and oil spill are just as effective and well financed today as they were in the aftermath of other spills. One of the most common tactics used in environmental disasters is to cast the blame on the opposition. Commonly, environmentalists are depicted as being responsible for the disasters. It was not long after the BP Deepwater tragedy that Sarah Palin led the charge in blaming the environmentalists: "Extreme deep water drilling is not the preferred choice to meet our coun-try's energy needs, but your protests and lawsuits and lies about onshore and shallow water drilling have locked up safer areas. It's catching up with you. The tragic, unprecedented deep water Gulf oil spill proves it" (Sheppard, June 2, 2010b).

Palin's allegation was picked up by every major media outlet in the nation. This was a curious statement considering that an American Petroleum Institute brochure, "Offshore Access" (2009), boldly states that "[t]he deepwater areas of the Gulf of Mexico represent a proving ground for technology—and a key component of America's energy future." The same page quotes a 2009 MMS interim report (May 2009) that states that 70 percent of all the oil and 36 percent of all the natural gas produced in the Gulf of Mexico is found in water deeper than 1,000 feet. In fact, the MMS report actually states, "Deep water has continued to be a very important part of the total GOM [Gulf of Mexico] production"(2009: 1). Thus, the oil industry was not forced to drill in deep waters by envi-ronmentalists; they were eager to do so because that is where most of the Gulf oil is located. In 2010, the top ten environmental organizations spent a total of $5,495,200 lobbying Congress. In the same year, the top oil and gas companies spent $38,178,838 (LaRussa, April 30, 2010). It is difficult to believe that environmentalists succeeded in shoving the oil and gas industry offshore against their will.

The *Nation* (Jones, March 1, 2010) has reported that scores of pun-dits, posing as disinterested experts appearing on cable news networks, are actually undisclosed corporate lobbyists and public relations offi-cials. These people are paid to manage corporate images and frame their analyses of current events in ways that distort the public narrative. This kind of "covert, corporate influence peddling," as characterized by Jones, demonstrates how press agents who are on the payroll of corporations can frame a story without revealing their bias. According to this reporter's account, "since 2007 at least 75 lobbyists, public relations representatives and corporate officials" appeared on cable network news (Jones, March 1, 2010).

The article quotes Jeff Cohen, a former MSNBC employee, who explained how these spokespeople appeared without any mention of their actual role as influence peddlers. When Cohen was asked by Jones how such individuals can appear on cable news without being introduced as lobbyists, Cohen replied that "these regulars get introduced the way they want to get introduced. This is the key: Gephardt [who now represents pharmaceutical and health insurance clients] will always be the former majority leader of the House. Period. These guys won't be identified by what they do now but instead what their position was years or decades ago."

Journalism professor Jay Rosen said in the same article, "Why are these people on at all? It goes without saying that individuals and communities victimized by the practices of corporations or government polices lack the resources and ability to gain access to the media and present counter narratives to the PR spin of large, vested interests in our society."

One example, not mentioned by Jones, is that of Patrick Michaels, who is described by the *New York Times* as a "climatologist who has long faulted evidence pointing to human-driven warming" (Morano, July 14, 2003). Michaels is a senior fellow in the conservative Cato Institute and has made such statements on NBC, CNN, and FOX, but the networks have failed to mention his affiliation. According to Sourcewatch, Michaels has received thousands of dollars over the years from fossil fuel industries (Hart, July, 2010: 7). This example is particularly striking since this is one of those rare instances where someone is framing global climate change, which is undoubtedly a disaster in the making, in advance of its full-blown arrival.

Of course, Michaels is not the only spokesperson who has attempted to become embroiled in the climate change controversy. In the wake of the TVA ash spill and the controversy surrounding TVA's handling of the spill, the TVA hired David Mould as the head of their public relations operation, in an effort to establish credibility for the agency (*Tennessean*, January 17, 2010). Mould was not on the job long before controversy of his appointment erupted over his alleged performance in his former job. As a former political appointee under the George W. Bush administration and, before that, a public relations spokesperson for the energy industry, Mould became embroiled in controversy at NASA over the alleged censorship of NASA scientists' statements on climate change. The NASA inspector general's report on the controversy implicated NASA's Office of Public Affairs and stated, "News releases in the area of climate change suffered from inaccuracy, factual insufficiency and scientific dilution." The report also revealed that the public affairs office also "managed the topic of climate change in a manner that reduced, marginalized, or mischaracterized climate change science." The inspector general's report also states:

> According to present and former career Public Affairs officers at NASA Headquarters and Field Centers that we interviewed, the NASA Headquarters Office of Public Affairs processed all media products that discussed "climate change" (or a variant thereof) in a unique manner during the pre-election period of the fall of 2004 through the spring of 2006. Describing the review process for climate change media products as extremely stressful, and heavy handed, it was their collective belief that there was an "air of political interference" and a desire by the political appointees in the NASA Headquarters of Public Affairs to support the Administration by reducing the amount or toning down the impact of climate change research disseminated to the public. (Inspector General's report June 2, 2008:18)

Although Mould is mentioned several times in the report, he denies having censored scientists. Setting aside any allegations of Mould's performance at NASA, it is notable that the TVA, which was desperately seeking to improve its tarnished image and project a new image of transparency, should choose a person like Mould, who has political baggage involving allegations of censorship and a lack of transparency. The appointment says more about TVA judgment than it does about Mould.

Robert McChesney and John Nichols (2010) provide us with an even more sobering analysis of the role of public relations in shaping the news. Press releases and press packets have played an instrumental role in shaping news since the 1970s. According to their account and that of Stauber and Rampton (1995), 40–50 percent of our news has been influenced by press releases. According to such reports, even in the days of a more robust news media, few, if any, of these press releases were investigated or edited before publication or broadcast. With the weakening of the press and the growing power of corporations, the situation has worsened. With fewer editors, smaller newsrooms, and smaller budgets, public relations firms are better situated to influence the content of the news. Former journalists are now better positioned to produce PR *news* in a way that McChesney and Nichols describe "as never being recognized for what it is" (2010: 47).

As news organizations shrink, more journalists are finding employment in public relations. In fact, McChesney and Nichols report that the number of PR specialists working with the news media (editors, reporters, announcers, etc.) has increased dramatically in the last three decades. While they cautiously admit that not all PR specialists work in the realm of shaping the news, they do state that such a dramatic change is indicative of a significant shift in the balance of power. No doubt as newsrooms continue to shrink, the balance of power will become even more asymmetrical. Such a shift does, of course, make it all the more difficult for individuals and community organizations to have their *unofficial* accounts heard. The ability of corporate powers to manufacture uncertainty and place their version of events into mainstream news frames makes the contestation over "what" and "why" a disaster has occurred challenging.

There are, of course, other ways in which news frames can be manipulated. Wikipedia, which is touted as a democratic, online way of providing information, is vulnerable to distortion and misinformation. According to Philip Coppens of *Nexus Magazine*, WikiScanner has discovered that a Dow Chemical Company (which now owns Union Carbide) computer was used to delete a section of the Bhopal disaster Wikipedia entry (January 20, 2008). The same scanner found that Exxon-Mobil was "linked to sweeping changes" in the Wiki entry on the *Exxon-Valdez* oil spill. An onsite allegation that Exxon, "[h]as not yet paid the US $5 billion in spill damages it owes to the 32,000 Alaskan fishermen," was, according to Coppens, changed to report that Exxon had paid the compensatory damages. One of ExxonMobil's IP addresses was linked to the edited changes. The BP web site makes promises, not unlike the promises made by Alyeska and Exxon in the wake of the *Exxon-Valdez* spill. Many of these promises cannot be realistically fulfilled, including cleaning the beaches to pre-spill condition.

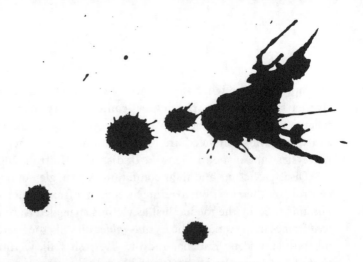

8. Contested Knowledge

*It is in the context of confrontation—when persons negotiate
their social universe and enter into discourse about it—that
the character of that system is revealed.*

(Comaroff and Roberts Rules and Process 1981: 249)

In our highly professionalized culture, the public debate over controversial topics is overwhelmed by privileged arguments. Experts attempt to exclude public knowledge from such ongoing controversies in order to keep knowledge "pure" (Tsing, 2005). Civil disputes, whether they are about the disposal of toxic waste, potential harm to public health, or other highly charged issues like those outlined in the preceding case studies, are seen as being fit only for expert debate. Lay questions, objections, and attempts to resolve uncertainty are often dismissed as uninformed, lacking in scientific vigor, irrational, and at times, almost hysteric. One woman whose life had been changed by the TVA ash spill recalled an exchange with a TVA official who avoided answering her questions and dismissed her reasoning. In response, she said, "Why do you treat us as stupid, why do you reject our arguments while upholding yours as the only reasonable ones?" (Enhorn, May, 2009). This frustration typifies the kind of rejection and frustration many disaster victims suffer in contesting official versions of reality.

Laypeople are made to feel as if they are inadequate arbitrators of uncertainty while scientists, corporations, and government agencies attempt to gain control over calamity. The determination to define and control the distribution and interpretation of knowledge is an attempt to define what is normative while excluding other alternative narratives. To a large degree, contemporary discourse governs what can and cannot be said about uncertainty. Thus, those voices heard and those silenced are of paramount importance in the controversies that ensue in the wake of disasters (Button, 2002).

Disaster victims and their communities struggle not only to regain control over their lives but to refute the *objective* frames offered by experts and mirrored by the media. Just as victims struggled to refute *normalized* frames in the wake of the *Exxon-Valdez* oil spill, the Shetland Islands oil spill, Hurricane Katrina, and the TVA ash spill, victims struggled to do likewise in the aftermath of Bhopal (Shrivastrava, 1987; Fortun, 2001), Seveso (Fuller, 1977), Love Canal (Gibbs, 1982; Levine, 1982), groundwater contamination in Legler, New Jersey (Edelstein, 1988), the Santa Barbara oil spill (Easton, 1972; Molotch, 1972), PBB poisoning in Michigan (Eggington, 1980), the nuclear accident at Three Mile Island (Moss and Sills, 1981), toxic contamination in Woburn, Massachusetts (Di Perna, 1985; Brown and Mikkelsen, 1990; Button, 2002), asbestos contamination in Libby, Montana (Peacock, 2003; Schneider and McCumber, 2004), Chernobyl (Medvedev, 1989), and radioactive fallout in the Marshall Islands (Johnston and Barker, 2008).

Attempts to silence lay voices in instances like these undermine alternative discursive frames of explanation and preclude analyses of catastrophes in a way that would fully allow unpacking the politically powerful world of social relations in which catastrophes and the calamity that follow are grounded. As noted in the introduction, science and technology are often implicated in many disasters, whether perceived as natural or unnatural. Thus, people, when faced with uncertainty, often instinctually turn to science and technology for answers. Some, if not many, people are dissatisfied with the perceived inability of science to answers questions and resolve uncertainties in a timely manner. Many people become disillusioned when they realize that science is often incapable of providing adequate explanations or true reassurance. Citizens have also become extremely distrustful of science and technology as well as the manipulation of these disciplines in the hands of both corporations and government agencies; they increasingly question the legitimacy of science and are wary of the adequacy of scientific approaches to prevent, mitigate, and recover from disasters. At times, science and engineering are seen as being the cause of catastrophic events such as in the case of the nuclear accidents at Three Mile Island and Chernobyl, the toxic contamination of thousands of Superfund sites within the United States, the chemical explosion in Bhopal, and even the collapse of the levees during Hurricane Katrina. As a result, many individuals view science as incoherent or too narrowly focused to undertake a holistic approach to solve their dilemmas or address the ontological challenges that they face in a time of calamity.

Some people regard the singular vision of experts as too narrow to deal with the multidimensional uncertainty that disaster victims face. Others simply become distrustful of experts and of even science itself. Others, however, attempt to become scientifically informed and become experts in their own right. Scholars have dealt at length with the contestation between laypeople and experts (Brown and Mikkelsen, 1990; Brown, 2009; Crouch and Kroll-Smith, 2000; Button, 2002). Other scholars have dealt with the process by which laypeople develop their own citizen's science (Brown and Mikkelsen, 1990; Giddens, 1991; 1994; Beck, 1992; 1995; Brown, 2009; Crouch and Kroll-Smith, 2000, Button, 2002; Button and Peterson, 2009) as well as the contestation between experts over scientific controversies (Rampton and Stauber, 2001; Davis, 2009; McGarity and Wagner, 2008; Michaels, 2008; Cranor, 2006).

When Experts Are Directly Affected by Disaster

Contestation can occur between experts whose lives are directly affected by disaster and nonaffected experts. In some cases, these affected residents are physicians, nurses, scientists, technicians, and attorneys. As experts, they use their knowledge to challenge other experts whose explanations are perceived as untrustworthy, incomplete, or false. In some cases, the experience of affected experts challenging other experts results in their viewing their profession and/or *rational* culture and its governmental infrastructure in a disarmingly new light. For example, in the case of the *Exxon-Valdez* oil spill, one expert who had long revered rational knowledge and science became skeptical of a scientific/technological approach that was inadequately prepared to respond to the challenges at hand. Moreover, she became distrustful of the manipulation of hard science to meet the demands of vested interests.

In each of the case studies above, a familiar pattern emerged. Uncertainties revolved around three primary areas of concern: effective remedial cleanup of contaminated areas, the immediate and long-term impact of the disaster on both the environment and public health, and just compensation for loss to both the environment and community residents. In all cases, both the corporate entities involved, and at least some of the government agencies involved, were perceived to have manipulated or distorted the facts in a manner that privileged the polluters and disenfranchised the affected population. Disaster victims and the general public struggled to obtain credible sources of information in an attempt to make sense of the disaster and to assign meaning and value as well as blame and responsibility.

A Multitude of Uncertainties

Multiple uncertainties haunted communities afflicted by disaster and left in a state of calamity. In Alaska, many communities were uncertain at first if the spill would reach their shores and affect their subsistence lifestyles and their commercial fishing waters. In tandem with this uncertainty was the uncertainty of how the spill would affect the livelihood of the commercial fishing communities and tourism. In the case of the Alaskan Native communities, uncertainty centered on not only the threat to their subsistence activities but to the preservation of their culture, already damaged under the colonial influence of first the Russians and then the Americans. In addition to these immediate concerns, there was considerable uncertainty about the long-term biological impact of the spill on the marine environment. The potentially adverse health impacts resulting from the toxic pollution of marine resources added another dimension of uncertainty to people's lives as did the occupational health concerns over the working conditions of the cleanup crews. Uncertainties abounded over the appropriate and most effective, least harmful, cleanup methods. Additional uncertainties centered on the most appropriate and humane way to rescue and rehabilitate birds and otters.

Halfway around the world, four years later, at yet another 60-degrees-north-latitude location, similar concerns were shared by many of the residents of the Shetland Islands. Foremost in many minds were the uncertainties over the impact of the spill on the salmon industry, crops, local livestock, the toxic contamination of houses from crude oil and dispersants, and the health of the residents of Donross Parish. This North Sea community was tremendously uncertain about the truthfulness of the local and national governments' statements about the impact of the spill. Everyone knew someone in this small community who had witnessed events that called into question the credibility of the government.

Many years later, in another coastal community along the U.S. Gulf Coast, residents faced a host of uncertainties that ranged from the loss of employment and housing, the separation of family members, and the ability to receive financial aid. Most central to our immediate analysis, they also faced uncertainties about the potential harmful impact of the flushing out of several Superfund sites, the failure of the levees, the toxic trailers in which they were forced to live for extended periods of time, the adverse health response many trailer residents and their families endured, and the potential health effects of crude oil and mold on the houses, property, schools, and public institutions in St. Bernard Parish.

Three years after Hurricane Katrina came ashore, the residents of Roane County, Tennessee, faced similar uncertainties. Residents living adjacent to the fossil fuel plant or downstream faced a litany of uncertainties ranging from the possible toxic contamination of their homes, their property, their wells, and the riverfront. A paramount uncertainty for many in the surrounding community was not only the contamination of a major river system, but the immediate and possible long-term health effects of exposure to hazardous fly ash waste. As one resident expressed, "Every time in the future I get a cold or flu I am going to wonder if I am suffering from the adverse effects of toxic contamination." Finally, many of these same uncertainties would eventually be faced by tens of thousands of people along the Gulf Coast, as the BP spill tragedy unfolds.

Downplaying Disaster

The controversies and uncertainties took on a litigious nature both in and out of the courtroom. A common theme surrounding the claims for compensation often involves who is rightfully entitled to compensation. In turn, this decision often depended on the ability of afflicted plantiffs to have their stories heard and legitimated. In the case of the *Exxon-Valdez* oil spill, legal disputes involving fishermen, Alaskan Natives, entrepreneurs, municipalities, and boroughs were litigated for many years. The bulk of the claims were only settled on the eve of the twentieth anniversary of the spill. This delay is testament to the enormous resources of the world's most profitable corporation to prolong litigation and wear down plaintiffs and their financial and legal resources. In later pages, we will examine in depth the litigious challenges plaintiffs face in the aftermath of disaster.

One uncertainty that has been debated for decades is what actually, if anything, transpired. Characteristic of most disasters is the perduring controversy over whether a hazardous event or a disaster actually occurred. Debate still persists as to whether an *incident*, an *event*, a *non-event*, or a *disaster* occurred. Thus, on the anniversary of events decades old—Love Canal, Three Mile Island, and Times Beach—there are renewed debates as to whether something of significance actually happened. When a disaster is acknowledged, the degree and magnitude of the event is debated. Over the persistence of disaster victims and their communities, the *official narratives* often endure while the narratives of victims are, at best, relegated to the footnotes of history.

In the wake of the *Exxon-Valdez* oil spill, one Exxon spokesperson emphatically stated, "I would not call it [the oil spill] a disaster." The recent thirtieth anniversary of the nuclear accident at Three Mile Island rejuvenated just such a debate. In the immediate aftermath of the TVA ash spill at the Kingston, Tennessee, fossil fuel plant, TVA CEO Tom Kilgore refused to label the event a disaster and referred to it as an "incident." In other instances, such as the chemical explosion in Bhopal, India, there is no denial that a catastrophe actually occurred, but a debate (as there often is in regard to disasters) over the magnitude and severity of the event is ongoing. Because of the quick manner in which the Indian government disposed of bodies in the Bhopal disaster, the actual number of lives lost is still debated. Union Carbide insisted that the disaster was not the result of negligence or technological failure, but sabotage, a theory that Union Carbide and others have touted for years (D'Silva, 2008). This theory has never been independently verified. An investigation by the *New York Times* concluded that the accident occurred because of a series of failures and not an intentional human act as Union Carbide has maintained. Nevertheless, many in the chemical industry still maintain that the catastrophe was not Union Carbide's fault but an act of sabotage.

In the final chapter, we will revisit the attempt of a corporation to downplay the consequences of a disaster, especially in regard to the dispute over the existence and size of undersea plumes in the case of the BP Deepwater spill. The specificities of each disaster, such as Love Canal, Three Mile Island, Woburn, Libby, Bhopal, Chernobyl, and countless other disasters vary from one another as do they from other disasters. Although many disasters are either less known or remain unknown in the public's imagination, there are common denominators that undergird all of these events. Residents in the affected communities did not possess the tremendous resources available to the corporate and governmental entities involved. The dogmatization of knowledge provided corporate and government experts a distinct advantage that allowed them to defend *official* versions of reality and refute counter-narratives.

Knowledge Contested: Information Withheld

In the case studies offered above, there are numerous examples of knowledgeable experts who find their lives directly affected by disasters and the uncertainties that follow. In the case of the *Exxon-Valdez* oil spill, foremost among these examples is Riki Ott, of Cordova, Alaska, a marine biologist and fisherwomen who has been actively engaged in such controversies for the last two decades, as has Rick Steiner, formerly of Cordova and now at the University of Alaska at Fairbanks. Another example, of course, is Dr. Rowlands who, along with several Indian scientists and physicians, became involved and remains involved in seeking justice for the victims and their families.

A more recent example is the TVA ash spill in Roane County, Tennesssee. Because of the presence of the TVA in eastern Tennessee, the nearby Oak Ridge National Laboratories, and Department of Energy regional offices, many people whose lives were affected by the ash spill possess a wide range of knowledge pertinent to the uncertainties surrounding the spill. Included among these knowledgeable citizens are former TVA engineers, technicians, and other employees. Despite the expertise of these local residents, the TVA, state, and federal agencies more often than not demonstrated a bias that privileged the official transcripts and sought to dismiss the *hidden transcripts* of experts contesting the TVA's denials. The framing process that ensued constructed and reconstructed meaning and value in a selective manner that privileged some political agendas and dismissed others, specifically those who contested TVA's downplay of the disaster and its potential environmental and human consequences.

In some instances, conflict between experts occurs when communities seek outside residents to assist them in their struggle. Love Canal residents sought the assistance of a number of scientists, one of whom was Dr. Beverly Paigen, a biologist and cancer researcher working for the New York State Department of Health. Paigen played a central and crucial role in the controversy and faced severe repercussions as a result of her involvement (Levine, 1982). A more contemporary example is Appalachian State University biologist, Shea Tuberty, whose water sampling test results are at odds with those of the TVA. Even more recently, several experts, including Wilma Subra and others, figure prominently in the recent catastrophic events in the Gulf of Mexico.

Another more dramatic example is Hugh Kaufman who was critical of the EPA's handling of the Murphy Oil spill. He was also an outspoken critic on Love Canal and Times Beach. In the 1980s, he was a whistleblower whose outspokenness resulted in the resignation of Anne Burford, the director of the EPA under President Reagan. Burford was caught up in a controversy alleging organized crime's illegal disposal of waste under an illegal contract that allegedly abused taxpayer money. Congress claimed that the EPA under her direction was involved in misusing $1.6 billion dollars in toxic waste Superfund site money. She refused to hand over records pertaining to the case to a Congressional investigation and was cited for contempt (Stauber and Rampton, 1995: 117–18).

The White House's ordering the EPA to delete information about the possible health effects of 9/11 were also highly criticized by Kaufman. In an interview on "Democracy Now," Kaufman, then senior engineer at the EPA and former chief investigator for the EPA ombudsman, stated:

> Basically, the report again documents what the ombudsman had documented over a year and a half ago: that the Environmental Protection Agency did not follow their usual procedures in a major emergency such as when the World Trade Towers came down. And instead, told the public that things were safe without even having the monitoring done to determine whether it's safe or not. The significance of that is that up until the World Trade Towers came down, the Environmental Protection Agency had procedures to follow to protect the public if a terrorist attack occurred and those procedures weren't followed and tens of thousands of people were put at and are at health risk because E.P.A. did not follow its usual procedures. The first responders, of course, received the major acute problems and of course we're seeing now that over half of those first responders are sick and the concern is that a large number of them will die very early because of cancer and other health effects because E.P.A. did not do its job properly. (August 12, 2003)

As we witnessed in the case of the *Exxon-Valdez* oil spill and the Shetland Islands spill, the ensuing controversies in the wake of Hurricane Katrina and in the TVA ash spill, there were, and continue to be, hotly contested disputes between the lay and expert communities. Both corporations and government agencies attempted to either withhold vital information or distort the facts in the acrimonious debates that followed between lay citizens and experts. In all cases, attempts to maintain an *official* account of the events resulted in amplified uncertainty.

In the chapters that follow, we continue to investigate how the professional domains of the media, public relations, law, science, the corporate world, and government agencies perpetuate uncertainty. The asymmetrical distribution of uncertainty underscores exactly how those who are most susceptible to the inflictions of uncertainty are relatively powerless in contrast to the manufacturers of uncertainty.

9. The Production of Uncertainty

Industry has learned that debating science is much easier and more effective than debating policy.

(David Michaels, *Doubt Is Their Product,* 2009: xi, emphasis added)

Voters believe that there is no consensus about global warming within the scientific community. Should the public come to believe that the scientific issues are settled, their views about global warming will change accordingly. Therefore, you need to continue to make the lack of scientific uncertainty a primary issue in the debate.

(Frank Luntz, Republican consultant, emphasis original, in Michaels, 2009: x 1)

In the foregoing disasters, some of the uncertainties were perceived as being generated by the corporations involved. In the case of the *Exxon-Valdez* oil spill, many thought Exxon's denial of the possibility of the spill leaving Prince William Sound as an outright denial of the facts, a denial that was confirmed in the minds of many when the oil spread in the exact manner local residents had predicted. The uncertainty over the possible short- and long-term health effects of the cleanup were also called into question by many people who felt that existing occupational health and safety data were being downplayed by Exxon and its subcontractors in order to obfuscate the problem. In later pages, we will reexamine this problem in our discussion of the BP Deepwater spill.

Similar perceptions were held about the conflicting data regarding the use of the dispersant Inipol. Both within and beyond the biological community, there was also uncertainty about the methods and procedures of the otter rescue program as well as questions of whether the program could achieve its stated aims of saving the otters. Some perceived the effort as a public relations stunt hiding beneath the façade of scientific research and humanitarian relief efforts. People were concerned that Exxon was generating misinformation in a variety of ways, including deceiving the public by manipulating the media, as in the case of the cleanup effort on the outer coast of the Kenai Peninsula. Finally, many community residents felt that, given Exxon's status and influence, having their voices heard over the *official* narrative of one of the world's largest corporations was difficult.

In the wake of Hurricane Katrina in St. Bernard Parish, residents worried about public health. Some of these concerns were generated by the previous polluting practices of refineries in their midst as well as by Murphy Oil. These concerns were exacerbated by the oil spill and Murphy Oil's questionable testing procedures as well as the testing and reporting practices of LDEQ and the EPA. The behavior of the company and of state and federal agencies created an air of uncertainty that made it difficult for residents to determine if reinhabiting the community was safe.

In the case of the TVA ash spill, uncertainty was produced in a multitude of ways including the restricted access to both the spill site and to test results. The very attempts of the TVA to downplay the spill and deny the disaster resulted in multiple productions of uncertainty. As we shall see in the final chapter, BP will take a similar approach in their response to the Deepwater spill, attempting to restrict access to the oiled areas and withholding valuable data.

Finally, in the Shetland Islands, many residents were also concerned about the employment of dispersants, especially close to local communities and the salmon fisheries. These uncertainties were reinforced when news was revealed that at least one of the dispersant products was a banned substance. Concerns about the health and safety both of livestock and humans also figured prominently in disputes following the spill. Frustration increased when the international press corps departed from the island leaving the community with few outlets through which to tell their story to the world. These concerns were centered almost exclusively on government policies and practices and not the behavior of corporations. A discussion of these concerns will be pursued in the following chapter on the role of government agencies in the bending of science and what I refer to as the use of "flexible" knowledge to protect their interests and, in many cases, the interests of corporate polluters.

Government agencies also contributed to the climate of uncertainty and sometimes either downplayed concerns in an effort to reassure the public or acted in tandem with powerful corporate interests to create an *official* narrative that overwhelmed the voices of citizens. These concerns about government behavior will be examined later. First, we will focus exclusively on the corporate production of doubt and uncertainty.

Science and Uncertainty

Science is as politically inflected as all other disciplines. As Stephens (2002) has rightfully observed, politics plays a formative role in deciding what constitutes "facts" in science. Contrary to its status in our culture as being neutral, impartial, and the ultimate arbitrator of facts, science is grounded in the political world of our everyday reality (Harding, 1991). Like all disciplines, science is more than politics, but it cannot evade the ideological currents in our culture any more than other sectors of society. This fact plays a decisive role in how the uses of science can obstruct disaster victims from having their case heard.

The dominant tactic in creating uncertainty and casting doubt on competing narratives is the subtle manipulation of science. While many citizens in the wake of disaster call into question the exalted status of science, some corporations seek to undermine the integrity of science by calling for a scientific approach that is ostensibly more rigorous than one predicated on principles of absolute of certainty.

In the United States, an ideological offensive was launched to argue that litigation and regulatory controls should not move forward until scientific analysis is capable of fully resolving scientific uncertainty. Even with the strictest practice of science, this demand can seldom, if ever, be met. Such a regimen dismisses most reputable scientific hypotheses that are commonly used in public and environmental health policy. More important, such an approach promotes the hidden political agenda that is the driving force behind these demands for *sound* science. It effectively forestalls regulation and protects the industry from expensive tort cases (Krimsky, 2006).

This attempted reshaping of science is somewhat curious give the fact that, as one scientist argues, science itself deals not so much in absolute certainties as in the weight of evidence. David Michaels, an epidemiologist and former assistant secretary of energy for Environment, Safety and Health and current assistant secretary of labor for Occupational Health, argues in his critique of the corporate production of doubt, that for scientists, proof of absolute certainty is almost never obtainable. Scientists strive to protect public health and the environment and recognize that it is not necessary to have truth beyond a reasonable doubt.

According to Michaels, corporate scientists have increasingly manufactured and magnified uncertainty in order to confuse policymakers and the public in the pursuit of the production of doubt. He contends that, "polluters and manufacturers of dangerous products tout 'sound science,' but what they are promoting just sounds like science but isn't" (Michaels, 2008: xi). The author contends that "emphasizing uncertainty on behalf of big business has become a big business in itself" (as quoted in McGarity and Wagner, 2009: 148). While many analysts, such as Michaels, contend that such practices are increasing at a rapid pace, the idea of manipulating and magnifying science in the interest of industry is not new. However, its practice and the sophistication of its tactics are on the rise.

The award-winning historians Gerald Markowitz and David Rosner have provided in-depth accounts of how corporate and government exchanges have for decades manipulated and distorted scientific knowledge (2002). In their book, *Deceit and Denial*, they provide an extensive history and examination of such practices (2002) with exhaustive accounts dating back almost a century. Using archival evidence and legal documents, the authors document a spiraling growth of strategies and tactics corporate powers have employed in order to avoid regulation and legal and moral responsibility. In calling into account corporate responsibility and sometimes government collusion in achieving these goals, they ask how those who bear a disproportionate share of the risk generated by these practices can effectively express their outrage and construct alternative realities apart from unbridled corporate abuse. In arguing why the public should have access to information that allows them to make informed decisions that affect their lives, they document how corporations "hide and obfuscate" information about toxics and how unchecked corporate practices prevent the public from making informed choices and decisions about safeguarding their health and communities. Their study documents in great detail how the lead, vinyl, tobacco, automotive, asbestos, and nuclear power industries have manipulated scientific research to deceive the public and, at times, the government. Their research provides a magnificent historical backdrop for those who call into question and criticize contemporary corporate attempts to distort science.

Michaels, Rosner, and Markowitz are not the only authors to focus on the increasing tendency, as legal scholars McGarity and Wagner call it, to "bend" science and create what they term "the antithesis of true science" (2008). In their opinion, not all science is equal given the deceptive practice of industry to bend scientific knowledge. The authors see such practices as pervasive in today's world. They are disturbed that research is too often "manipulated to advance economic or ideological ends." They argue that science is literally under attack and that very sophisticated practices are now being employed to *co-opt* science that informs public health and environmental policy.

In their detailed study of the corruption of science, they acknowledge, as Markowitz and Rosner do, that the birth of such practices has deep historical roots in American history. They argue, as does Michaels, that the attenuation of science has accelerated in recent times. Part of their concern is that too few resources exist in the government and the media to monitor and regulate "outsider-induced scientific distortions." They are particularly concerned that legal analysts and the courts mistakenly believe that internal monitoring and practices within the profession of science alone are sufficient to ward off such distortions from undermining legal and regulatory arenas (McGarity and Wagner, 2008). They argue, however, that professional oversight within science is incapable of doing so. McGarity and Wagner argue that the risk of the contamination of good science is largely denied by many attorneys and judges because of their over-reliance on a scientific filtering process and regulatory policy-makers, people, and institutions that ignore the fact that the system has become handicapped and broken by ideological and economic forces (2008). Moreover, they argue that when special interests are caught skewing scientific evidence, too often it is viewed as an exception and not perceived as a systemic failure of both science and the legal system to restrain such practices (McGarity and Wagner, 2008).

Corporate Tactics and Court Cases

McGarity and Wagner outline the techniques in which corporate advocates bend science in order to "manipulate, undermine, suppress or downplay scientific outcome." Here is a brief synopsis of these tactics:

Shaping Science: which they describe as commissioning research that has the stated purpose of producing a particular outcome. In many cases the sponsors of the research can frame the research question, even design the research protocols and have a considerable influence on the study's outcome often in ways that are subtle and difficult to detect.

Hiding Science: described as a "second-best" effort to suppress unwelcome findings. This is a low cost effort that has the potential to yield great benefit. Of course it is only an option if a company or a government agency has sole possession of findings.

Attacking Science: described as creating "illegitimate" attacks on research that might be damaging to a vested interest. Typical tactics include challenging methodologies, data interpretations and review processes. Such baseless criticisms can appear more legitimate if made by a number of seemingly different parties.

Harassing Scientists: basically a variety of tactics that can be employed to make life difficult for scientists whose work appears threatening. Attacks can range from attacking the integrity of the researcher, alleging unsupported allegations of misconduct, requesting a time-consuming, overwhelming amount of data, harassing subpoenas and depositions that can deter researchers from continuing with their research agendas.

Packaging Science: This can entail commissioning review articles that summarize research evidence in favor of the sponsor. Cherry-picked experts can knowingly or unknowingly serve this purpose. Another tactic is to stack a conference research panel in favor of the client. Such tactics cannot only influence scientific discourse but once planted in the research literature can be used as evidence in a trial in order to influence judges and juries.

Spinning Science: Public relations experts can re-interpret studies in ways that can suggest the studies are flawed, tentative and require more research, or characterize the research as junk science. In the hands of a complacent or ignorant press, or a press that doesn't have the time, resources or capacity to investigate these allegations these distortions can go unchallenged. (McGarity and Wagner, 2009: 38–40)

Social theorists have also studied how corporations attempt to influence public agendas and define social problems in ways that undermine the claim-making activities of those who may threaten their profits. These indirect forms of manipulation by powerful interests entail a number of tactics similar to those outlined by McGarity and Wagner, including redefining issues and events as non-issues. These subtle forms of diversionary reframing entail the revisioning of knowledge through various strategies and tactics (Molotch, 1972; Bachrach and Baratz, 1970; Renson, 1971; Schnailberg, 1994; Krogman, 1996; Freudenberg, 2005; Freudenberg, 2008; McCraight and Dunlop, 2003). Perhaps the classic case of such attempts to influence public opinion is the tobacco industry's attempt to argue that there was not enough science to regulate tobacco products (Warner, 1986; Glantz, et al., 1995; Hilts, May 7, 1994, 1996; Rampton and Stauber, 2001).

For our purposes, one of the most interesting examples is Exxon's attempt to bolster its appeal by publishing articles in prestigious academic journals and law reviews. Exxon intended the accomplishment of two purposes: 1) questioning the competency of juries to set punitive damages fairly and 2) making a case for the argument that such large punitive damages are bad for society. In this instance, Exxon was seeking to set aside a 1994 settlement of $5.4 billion in punitive damage that juries had awarded plaintiffs for the harm caused by the *Exxon-Valdez* oil spill. Exxon succeeded in hiring nine social scientists from a broad array of disciplines to write articles in support of their cause, many of which were published in peer review journals. Then, in 2002, Chicago University Press published the book *Punitive Damages: How Juries Decide* in which the contributing authors acknowledged the outside funding. The articles served to galvanize Exxon's claims. In its appeal to the Ninth Circuit Court, Exxon cited a number of the papers without mentioning the fact that they had funded (at an estimated cost of $1 million) and initiated the research in this book and other venues. Richard Lampert, a University of Michigan law professor, criticized the tactic saying, "It is very troublesome that work published as scholarship … is being vetted by lawyers" (Krimsky, January 16, 2007).

All the book contributors claimed they maintained intellectual control of their research. However, they admitted that officials from Exxon commented on the drafts and coordinated meetings among the authors. At least one of the authors refused funding and travel expenses (Zarembo, 2003). The Ninth Circuit Court returned the case to a lower court for a ruling. Determining whether the papers made a difference in the court's decision is hard to say, especially because many other legal issues pertaining to the case may have been more influential in the outcome, which ruled in favor of Exxon. Sociologist William Freudenburg has written a very candid and self-reflected article about his drafting a preliminary article for Exxon. Eventually, he and the company went their separate ways. Exxon was uncomfortable with his conclusion that transparency encourages responsible corporate behavior. In his article, he quotes an Exxon contact in a phone conversation about the corporation's intentions:

> Basically, what we were exploring is whether it is feasible to get something published in a respectable academic journal about what punitive damages do to society or how they are not a very good approach. Then in our appeal, we can cite the article and note that professor so and so said in this academic journal, preferably quite a prestigious one, that punitive damage awards don't make much sense. (Freudenburg, 2005: 14)

Freudenburg later drew his own conclusion about his interaction with Exxon: "The legal system and scientific method co-exist in a way that is hard on truth" (Liptak, November 24, 2008). Exxon's attempts to influence the court did not end there. In 2008, the case made it to the Supreme Court, which ruled that the appropriate punitive damages were around $500 million, not the $5.4 billion the jury had awarded. The decision was obviously a major setback for the plaintiffs. A *New York Times* article reviewing the decision (Liptak, November 24, 2008) stated that Justice Souter, who wrote the majority decision, stated in a footnote that, "because this research was funded in part by Exxon we decline to rely on it." Since then there has been much debate in legal circles about the potential influence of the funded research. The *Times* article quotes Terry N. Gardner, who coordinated the Exxon project and who had contacted Freudenburg on Souter's disclaimer: "My feeling was they seemed to have an obligation to say that. Yet the arguments the justices used in part reflected the conclusions of the study." This opinion was held by many of the plaintiffs as well.

A more recent example of the attempt of an oil company to influence public opinion surreptitiously is an article in *Atlantic Magazine* by freelance reporter Mary Cuddehe (Cuddehe August 2, 2010). Cuddehe says that Kroll, one of the largest private investigation agencies in the world, called her last February and invited her to pretend she was a journalist in order to act as a spy for Chevron. For several decades, Chevron had been accused by accused by Ecuadorian Amazon residents of oil drilling practices that have resulted in spilling more than 18 million gallons of oil and toxic waste in the Amazon Basin. Critics say that such practices have had a deleterious effect on nearby residents and the environment. The ongoing dispute with Chevron has resulted in a $27.3 billion lawsuit against the oil giant for harmful health effects on residents in the affected area. According to Cuddehe, Kroll, acting on behalf of Chevron, wanted her to pose as a journalist while investigating a health study that reportedly demonstrates that the oil companies' practices in Ecuador have caused high cancer rates among local residents. Chevron maintains the health study was "rigged" by the plaintiffs. After meeting with a Kroll representative, Cuddehe declined the offer and decided to write an expose instead (see also *Democracy Now,* August, 16, 2010).

There are, of course, legal considerations to consider as well when discussing the bending of science or the production of doubt and uncertainty. Within the constraints of this study, examining in its entirety the complex, jurisprudence considerations that can come into play in tort law is not possible. Carl Cranor has written an in-depth book on toxic torts that extensively investigates the topic (2006).

Cranor makes several salient observations that provide us with an essential understanding of the fundamental issues in this regard (2006). He outlines for the reader the basic tensions that exist between science and law that contribute to our understanding. First and foremost, he points out that time is a critical difference between these two disciplines. While the law imposes time constraints in terms of the statutes of limitation, science has different time constraints. For instance, a scientist must first obtain funding and research consent before formal research can begin. Although Cranor does not mention this, I hasten to add that those who pursue disaster research must often wait a protracted period of time before receiving either, especially funding, an impediment that makes the early documentation of important elements in the wake of disaster difficult. Cranor, however, is quick to point out that research that is required in legal settings may or may not conform to the time restrictions stipulated by law. In the case of toxic contamination, especially the effects of long-term toxic contamination, such restrictions are critical, given the fact that it may take a long time (i.e., several years) for the manifestation of symptoms related to toxic exposure. While it often takes considerable time for toxic exposure to demonstrate harm, it can take considerably longer to document the extent of the damage.

In such instances, this is not only a severe constraint on toxic court cases, but often stymies long-term health effect research as well. Longitudinal funding is difficult to obtain even when there is the scientific and political will to conduct the research. Cranor adds that this difficulty can also be crucial because many plaintiffs may not have sufficient funding to conduct such protracted research on their behalf. Obviously, corporations seldom face such constraints.

The toxic groundwater contamination in Woburn, as illustrated in Harr's book *A Civil Action*, provides us with a perfect example of the dilemma of time as a central consideration. Not long after the court case was settled, the EPA issued a report proving the chemical contamination in question did actually reach the wells in Woburn, just as the plaintiffs had alleged. Harr's comment aptly illustrates the problem with the time differential between the law and science: "On the face of it, the verdict appeared to stand for an example of how the adversary process and the rules and rituals of the courtroom obscure reality" (Harr, 1995: 456; also cited in Cranor, 2006: 216).

Cranor's third distinction between science and law, one that he calls the most significant, is that the two disciplines have different standards of proof. The fourth distinction that Cranor highlights is that scientists, in their approach to a problem of complexity, are more nuanced and sensitive to the complexity of their subject. Because courts need to make the law at least somewhat accessible to the public, they have a tendency to simplify.

Finally, the author's fifth point is of vital significance to our exploration of uncertainty. Cranor makes a point mentioned earlier in the text, which Michaels has also observed: Uncertainty is endemic to science, and given that this is the case, scientific claims are more reserved and cautious in the admission of what is known and unknown. They are, as Cranor suggests, much more comfortable with varying degrees of uncertainty. The notion of uncertainty is treated quite differently in U.S. jurisprudence. While there are rules for dealing with uncertainty, which as Cranor points out often deal with legal presumptions, burdens of proof, and standards of proof, there are constraints in instances where there is excessive uncertainty. The party with the burden of proof on the issue loses the case. Courts, as Cranor observes, "do not have the luxury of avoiding a decision in the face of uncertainty" (2006: 217). This burden, as Cranor astutely observes, "can lead to deliberate strategies, especially on behalf of the defense to emphasize or exaggerate the uncertainties." The reasons for which this is true have to do with what many refer to as, "the most important Supreme Court decision which you never heard about" (the *Daubert* decision) (Tallus, 2003).

The history of this decision is noteworthy, given our previous discussion about the use of publications to influence court decisions. It began with corporate funders, Peter Huber, and the conservative Manhattan Institute's goal of discrediting toxic tort cases filed against corporations. In a public relations campaign designed to weaken such cases, Peter Huber wrote a book, *Galileo's Revenge,* in which he attacked the science employed by plaintiffs, their attorneys, and, most important, their data and expert witnesses (1991). Huber labeled scientific evidence deployed in these often successful challenges to corporate practices as "junk science": "Junk science is the mirror image of real science, with much of the same form but with none of the substance... .It is a hodgepodge of biased data, spurious inference, and logical legerdemain... .[I]t is a catalog of every conceivable kind of error, data dredging, wishful thinking, truculent dogmatism, and, now and again, outright fraud" (1991).

This factually inaccurate, ideologically driven book became enormously influential in some corporate and legal circles. Its success elevated Huber to one of the most influential voices in the policy debate over tort reform. When the case of Daubert was remanded by the U.S. Supreme Court, the U.S. Court of Appeals for the Ninth Circuit played a crucial role in the Court's decision. Judge Alex Kozinski cited the book as a primary basis for writing his opinion on Daubert (Cheseboro, 1993; Tallus Institute, 2003).

The U.S. Supreme Court decision in *Daubert v. Merrell Dow Pharmaceutical, Incorporated* (1993) established tests for judges to employ in deciding if the science presented in a case is credible and admissible. It establishes judges as gatekeepers who can make pretrial decisions regarding what scientific evidence can be admissible. Among other points, the ruling rejects animal studies as relevant to human harm. This decision places undue and unrealistic expectations on the role of epidemiology in resolving uncertainty, by stating that epidemiological studies, if available, trump all other studies. This decision demonstrates a lack of true understanding of the constraints of epidemiology and the reasons for which animal testing has played such an important role (Davis, 2007: 235).

The decision also provides opposing sides in a case the pre-trial right to challenge the credibility of expert witnesses. It also provides judges with broad discretionary power to decide not only what is admissible as evidence, but also who can testify as an expert witness. In civil trials, as Michaels (2008: 162) points out, plaintiffs normally bear the burden of proof and, thus, must present supportive evidence. Usually, the defense makes the Daubert challenge as to the admissibility of evidence and expert witnesses.

If the judge dismisses the admissibility of the evidence, commonly, the judge will make a summary judgment in favor of the defendant and dismiss the case. In other words, the case will never go to trial, and the plaintiffs will lose their day in court. This practice places a tremendous burden on plaintiffs and often requires that they invest a considerable amount of money to establish their case. As Michaels (2008: 173) argues, the main beneficiaries of the ruling are the defendants because they are usually corporations with considerable financial and legal resources. They can afford expensive expert testimony and pay for researchers to find ways to cast uncertainty on the defense's claims.

It also opens the door for the possibility that the defendants in the case, seeking to win on evidentiary grounds prior to trial, will be tempted to, as Cranor, says "exploit uncertainty in the science, to exploit its high implicit standards of proof, or even create a misleading idea of the science needed for toxicity assessments" (2006: 349). If the defendants prevail, Cranor states, "the case is over." Cranor argues further that Daubert, "creates incentives for firms to distort scientific studies" and also "creates incentives to distort scientific literature" in this pretrial phase.

This ruling has raised considerable concern in many legal and public health quarters (Cranor, 2006; Davis, 2007; Michaels, 2008; McGarity and Wagner, 2008). Chief Justice Rehnquist, in a dissenting opinion, joined by Justice John Paul Stevens, questioned the wisdom of turning judges into "amateur scientists" (Michaels, 2008: 165). Other critics charge that the ruling turns judges into referees. Furthermore, it favors those parties that demand that human harm must be proved to have actually occurred before any claim of a casual connection between exposure and deleterious human effects can be made. Critics also allege that the ruling is based on fundamental misunderstandings of how science actually works as well as a lack of understanding of the constraints of epidemiology (Davis 2007: 323–24).

It is commonly accepted in the scientific community that, concerning toxic substances, epidemiological evidence, for the most part, does not exist. Since it is unethical to expose people to toxins in order to establish exposure and dosage levels, such evidence is hard to come by. Thus, instead of creating natural experiments, scientists have to use what data exists and wait for additional evidence to emerge (Tallus, 2003: 8).

As Davis and others argue, the courts' decision makes it mandatory that human harm has to occur in significant numbers, with stringent rules of documentation. Such a ruling ignores the fact that documenting actual harm can take years. Therefore, such a ruling severely retards attempts within the legal system to prevent harm to human health. Under such guidelines, how many individuals would have to be stricken and perhaps die before sufficient evidence could be presented to the court? This fact alone greatly diminishes the ability of court decisions to act in a protective manner.

Moreover, scientists like Michaels (2008: 165) argue that the Daubert ruling provides judges with "no philosophical tool to judge 'good science'," for, as he rightfully argues, in science there are no absolute criteria for "admitting the validity of scientific evidence." For example, the court's decision is predicated on the notion that absolute certainty in science is the norm. Michaels and others argue that certainty "is rarely an option in science." Uncertainty is the norm and not the exception (2008: 165). He and other scientists contend that scientists usually have to make judgments based on the weight of evidence rather than rely on certainty.

Ironically, by calling for an elevated standard of uncertainty, the decision provides, in some instances, a nearly perfect vehicle for the production of uncertainty. Davis perhaps best characterizes some of the problems with the court's decision: "The Daubert decision presumes that well done science is like painting by numbers, displaying universally agreed standards and methods to come up with clear facts" (2007: 324).

"Uncertainty can also be the recipe for regulatory analysis," argues Michaels (2008: 251). He provides two examples of how Daubert has inspired industry and trade groups to establish similar standards for regulatory hearings. Corporate lobbyists, at a hearing that examined OSHA's process for establishing health and safety standards, made the recommendation that Daubert be incorporated in OSHA's standards. Potentially just as harmful, the U.S. Chamber of Commerce in 2002 adopted an official position on scientific information in federal rule-making that was highly influenced by the Supreme Court's decision. They advocated an adoption of an executive order that would require all federal agencies to adopt the Daubert standards in the administrative rulemaking process (Michaels, 2008: 174). As Michaels quite rightfully asserts, such an approach "runs directly counter to the precautionary policies built into most health, safety and environmental statutes." McGarity and Wagner also express concern for applying the court ruling in the regulatory arena. They argue that such developments expend "the menu of tools available for bending science with no appropriate correlative benefit for improving health or environmental protection" (McGarity and Wagner, 2008: 26).

Krimsky has pointed out that the manner in which courts assess exposure to toxic hazards is at variance with the way regulatory agencies assess the risk (2003: 14; Tallus, 2003: 16). The EPA and OSHA depend on the weight of evidence, which is an accepted practice within the scientific community. That is to say, all evidence is considered in the assessment of exposure to toxic hazards rather than evaluating each piece of evidence in isolation. This approach creates a tremendous imbalance. The courts' assessment approach is biased in favor of the defendant and against the plaintiff (Tallus, 2003: 16). One cannot help but wonder how the Daubert decision will provide a decisive advantage to BP in the open litigation involving the Deepwater blowout.

The endless insistence on science as the ultimate arbitrator in legal and regulatory circles ignores the prevailing influence and consideration of other influences that also need to be considered. While in many ways it is imperative to pursue scientific answers to safeguard the public and the environment, it is wrongheaded and absurd to insist on absolute certainty before acting. As Davis has so astutely observed, "[I]f we always insist that we should do nothing until the damage is absolutely certain, then the only certainty is that we will cut short millions of lives and bring misery to millions of others" (2002: 379).

In examining the challenges and ongoing suffering of the victims of the chemical explosion in Bhopal, Veena Das makes several observations relevant to our discussion. The manner in which the medical knowledge was constructed on behalf of the defendant, Union Carbide, obstructed the ability of the plaintiffs to argue their case. The victims, Das observes, had to transfer their suffering into the language of science, a language that was highly constricted by both the State of India and the defendant's manipulation of the facts. Thereby, the victims were forced into a position that made them appear as though they "were responsible for their disease not being understood by modern science" (Das, 2000). While the specificities of the tragedy in Bhopal and the legal system in India are decidedly unique, the plight of the defendants described by Das is all too familiar.

10. Sequestered Knowledge

Every bureaucracy seeks to increase the superiority of the professionally informed by keeping knowledge and intention secret...In so far as it can it hides its knowledge and action from criticism.

(Max Weber, *Essays in Sociology*, 1952: 223)

In the contestation that follows in the wake of disaster, government agencies play a role equally decisive to that of the media, the law, and corporations in framing the event. How such agencies interpret, shape, and dispense with knowledge is crucial to the political process. The way in which knowledge is produced in an atmosphere of pervasive ambiguity and uncertainty is critical in the calamity that follows. How government agencies behave sometimes closely resembles the behavior of corporations. As Harvey points out, in a neoliberal state, it is often the "public sector that bears all the risk and the private sector that reaps all the profit" (2006: 26).

Too often the decisions government agencies present to the public are presented in an atmosphere of absolute certainty (Das, 2000). In an attempt to quell a problem, governments often offer the public a false sense of security. Like corporations, they are often guilty of denying the existence of a problem in an attempt to avoid public scrutiny. In cases such as those we have examined, this practice can result in the dilution of scientific integrity.

In this atmosphere of reassurance, toxic chemicals can be deemed innocent until proven guilty. Yet government responses often generate considerable distrust in the agencies' handling of future crises. For instance, the role of the U.S. EPA in concealing facts about airborne dangers in the wake of the 9/11 attacks on the Twin Towers (2001) raised serious doubts about its credibility in response to airborne hazards in the wake of Hurricane Katrina (2005) and the TVA ash spill (2008). Furthermore, it undermined the Agency's credibility during the BP Deepwater spill. In all three cases, suspicions were confirmed when it was revealed that proper precautions and transparency were also lacking as they were in the case of 9/11.

The Translocal Nature of Disaster

The controversy surrounding the federal government's handling of air quality issues in the aftermath of the September 11, 2001, attack on the World Trade Center in New York is not only illustrative of how knowledge is shaped in a time of calamity but also how disasters are sometimes imbricated in uncanny ways. The tragedy of 9/11 is a perfect example of how disasters are both situated in a society's history and are very often translocal. Although the 9/11 disaster was triggered by an attack from without, one of the potentially most harmful outcomes from that tragic day in 2001 stems from a previous chronic disaster that unfolded three decades before, in Libby, Montana.

In this small town in northwestern Montana, the W. R. Grace Company mined vermiculate containing high amounts of asbestos. In large part, the story of the disaster in this northern town is a story of how W. R. Grace, the state of Montana, and the federal government concealed the asbestos hazards from both the town and the nation.

The Grace vermiculate was the single largest source of asbestos in the world and sold around the globe in products ranging from insulation to crayons. Several hundred miners and their families died in Libby, and many more residents with asbestosis, mesothelioma, and lung cancer are likely to perish. Along with the thousands of workers and homeowners around the world who handled the material or were otherwise exposed, Libby, Montana, along with global climate change, may well be the ultimate examination of a translocal disaster.

Asbestos played a central role in the airborne contamination resulting from the attack in the heart of New York City. In order to prevent lightweight steel beams from melting in a fire, a W. R. Grace fireproofing material containing a vermiculate-gypsum plaster, high in tremolite asbestos, was sprayed on many of the beams in both towers. Estimates of the amount of asbestos used just in the towers range from 250 to 750 tons (Peacock, 2003: 156). All of this asbestos, along with tons of other hazardous chemicals, pulverized concrete, and heavy metals, were dispersed when the towers toppled. Moreover, the dust was found to be highly caustic with a pH level comparable to ammonia or, in worse cases, drain cleaner (Peacock, 2003; Schneider, 2004). This toxic dust spread in a plume over lower Manhattan and Brooklyn. The plume covered schools, hospitals, residences, office buildings, workplaces, businesses, and shops as well as firemen, policemen, social workers, and medical personnel at ground zero. The area surrounding the site was covered in a thick toxic dust that covered the outside and permeated the inside of buildings.

People in the adjacent area and as far away as Brooklyn complained of the "9/11 cough." Early symptoms included, sneezing, wheezing, asthma, respiratory difficulties, sore eyes and throat, weakness, and flu-like symptoms. Christine Todd Whitman, director of the EPA, reassured New Yorkers shortly after the attack: " Given the scope of the tragedy from last week, I am glad to report that their air is safe to breath [sic] and their water is safe to drink" (Preston, February 3, 2006).

Controversy over Whitman's statement and the EPA's onsite testing soon erupted. Scientists and officials inside and beyond the agency cringed when hearing Whitman's statement. Dr. Phil Landrigan, head of Mount Sinai Medical Center's Department of Community and Environmental Medicine—arguably, the most informed department of its kind about the hazards of asbestos in the nation—was concerned about the tons of asbestos clouds swirling around lower Manhattan. He was convinced the dust would cause serious health problems to the emergency workers on site and the tri-state residents exposed to the asbestos debris (Schneider and McCumber, 2004: 334). Experts around the country said that the risks were being grossly underestimated. While the EPA was saying that the detectable levels of asbestos were below the threshold of safety, many people within and beyond the federal government knew that there was no safe level of exposure. Theoretically, just one low level exposure could prove to be lethal (Schneider and McCumber, 2004: 345).

In August of 2003, the Office of Inspector General for the EPA issued a report highly critical of Whitman's press statement. The report revealed that the agency's statements to the press and public had been tampered with by the White House. According to the inspector general's findings, cautionary data had been deleted from the statements while reassuring information was added to give the impression that there was no threat.

When the EPA made a September 18 announcement that the air was "safe" to breathe it did not have sufficient data and analyses to make such a blanket statement. ... Furthermore, the White House Council on Environmental Quality influenced ... the information that EPA communicated to the public through its early press releases when it convinced EPA to add reassuring statements and delete cautionary ones. (U. S. EPA, September 26, 2003:17)

Among the deleted sentences was the following: "However, even at lower levels, EPA considers asbestos hazardous in this situation and will continue to monitor and sample for elevated levels of asbestos and work with the appropriate officials to ensure proper awareness."

Not only had the reassurances been made without scientific evidence to support such claims, the White House interfered with the EPA's original statement to the public. Deleted were EPA statements that asbestos levels in some areas tested by the agency were three times higher than national standards. The statement was amended to read, "slightly above the 1 percent trigger for defining asbestos material" (U. S. EPA, September 26, 2003:16). Furthermore, the report concluded that EPA's reassurance (made on September 16) that it was safe for New Yorkers to return to work on Wall Street was substituted for a statement with a very different message, one that stated that the air monitors failed to record dangerous samples. Statements were also deleted that indicated caution for high-risk populations (people with asthma and respiratory diseases, the elderly). Thus, these edits and others made by the White House Council gave New Yorkers a false sense of security.

James Connaughton, a Vice President Cheney political appointee, chair of the White House Council on Environmental Quality, and a former employee in the mining and asbestos industry, was identified as the man most responsible for the White House interference. Throughout his tenure on the Council, he played a central role in distorting science. In his first days in office, he successfully implemented Cheney's energy plan by making oil and gas drilling permits and leases on federal land easy to obtain. He also played a central role in the Bush administration's downplay of the threat of global climate change. Emails on the White House website indicate he was very active in working closely with Exxon-funded organizations to undermine the science of global warming (Bowen, 2008: 112).

Critics both within the EPA and beyond were critical of the testing methods that the EPA conducted. Rather than use the state-of-the-art approach employing the latest electron microscope technology and fiber counting protocols used by the best private labs in the country—many of which were frequently employed by the EPA to conduct such tests and ironically some of which were used for the City of New York—the EPA elected to employ equipment and methods that were far less accurate.

In fact, one EPA scientist claimed that the difference was crucial and "too important to be ignored if you really care about the health of the public" (Stranahan, January-February 2003). Cate Jenkins, a twenty-two-year veteran of the EPA and a senior chemist, claimed that testing methods utilizing the advanced technology found nine fibers for every one that the EPA detected (Schneider and McCumber, 2004: 341). In making public much of what the EPA was trying to suppress, Jenkins emerged as a whistle-blowing hero. According to Jenkins, her access to the data was just a chance mistake. Apparently, someone in the New York City Department of Health mistakenly transferred over the data to the New York State Department of Health website from which she was able to obtain the database to which she previously had, along with other EPA employees, been denied access (Markowitz and Rosner, 2002). Jenkins converted the facts that the EPA collected into PDF files and posted them on the internet.

Jenkins alleged that the "EPA falsified the data in a callous disregard for human life." According to Jenkins, "The die was cast. EPA [then] had to cover its tracks and claim that there was no hazard. After they did that, their only thought was protecting themselves from being sued. They displayed absolutely no respect for the health and safety of the public" (Markowitz and Rosner, 2002). Journalist Andrea Peacock reported that Carl Weis, an EPA toxicologist describing Jenkins's action, said, "[W]hat it has done is it's taken the public relations machine that EPA established and turned it on its ear. ... [W]hat was being said ... to the press ... was not consistent with the hard facts, the data" (Peacock 2003: 170).

The federal government also suppressed other data gathered on the asbestos risk, arguably the most valuable. The U.S. Geological Survey (USGS), in cooperation with the EPA Denver office, obtained incredibly accurate information about airborne contamination by utilizing a sensing unit that was originally designed for exploring dust on the surface of planets. USGS scientists and NASA pilots obtained permission to fly over Manhattan on three different occasions and accurately detected the high levels of toxic contamination including previously unknown and extremely high levels of pH. According to Schneider, the U.S. government would not allow the team to share the information with the public or other agencies (2004: 35). Schneider reports that even some of the political appointees in the EPA, FEMA, and Health and Human Services, "were outraged at being forbidden from sharing what they knew about the dangers that existed because of the dust" (2004: 337). Moreover, those agencies and professionals working at Ground Zero to protect the health of the workers and the public were not given this information. In fact, the USGS information on the contents of the toxic dust was not disclosed until February 10, 2002, when Schneider's story was broadcast by the networks. The seriousness of withholding this vital information is illustrated by two health officials that Schneider interviewed.

One, Carrie Loewenherz, an industrial hygienist for the New York Committee for Occupational Health and Safety, stated, "[T]hat was information we all should have had." The director of the Mount Sinai Center for Occupational and Environmental Medicine, Dr. Robin Herbert, who "was supposed to be in the loop," told Schneider, "There's a large segment of the population here whose physicians needed to know that information. … There is no justification for holding it. You don't conceal the information from those who need it" (Schneider and McCumber, 2004: 346–67). Withholding vital information that is of value to physicians treating their patients and public health officials trying to protect the public is reminiscent of the withholding of vital medical information in the wake of the chemical explosion in Bhopal, India (Shrivastava, 1987; Fortun, 2001; Lapierre and Moro, 2002).

The federal government's restrictions on sharing information is not unlike the behavior of the federal government when, in the wake of the anthrax attacks, it prohibited government agencies from sharing vital information from one another. Neither instance is an isolated case of this kind of coverup. In the aftermath of Hurricanes Katrina and Rita, several mid- and upper-level officials anonymously told me that Homeland Security withheld important information from them by blocking their computer access passwords.

Congressman Carolyn Maloney from New York City was among the many people who were disturbed by the EPA's behavior. She found it "extremely troubling" that, as a result of the Whitman's statement to the press, "people stayed [on site] longer than they should and went back to work without any respirators" (Markowitz and Rosner, 2002). Hugh Kaufman, an EPA investigator who, as we shall see in the next chapter, also became an outspoken critic during the BP Deepwater disaster, was also troubled by this neglect of workers' safety. "The people in the pit should have been wearing respirators. Many of them were ordered by their boss not to wear respirators because it might scare the public" (Markowitz and Rosner, 2002). In the same interview, Kaufman went on to say that "I have never seen the EPA or the government be as irresponsible as it's been in its response to 911" (Markowitz and Rosner, 2002).

The press largely ignored the story, with one exception. *New York Daily News* reporter Juan Gonzalez, in the face of considerable political pressure, took the lead in writing about the health hazards and cover-up. His stories were so impactive that Mayor Rudy Giuliani called a press conference and announced, "[T]he problems created ... are not health-threatening" (Hagey, April 17, 2007). The head of the New York City Partnership and the Chamber of Commerce also complained. They sent a letter to the newspaper saying Gonzalez's column was a "sick Halloween prank" (Hagey, April 17, 2007). A deputy mayor from Guiliani's office called the *Daily News* and complained.

While other editors at the *News* wanted to back away from Gonzalez's coverage, metropolitan editor, Richard T. Pienciak, established a four-member investigative team to delve deeper in to the story. Earlier in his career as an investigative reporter, Pienciak had covered the Three Mile Island Nuclear Accident and had experience with government agencies withholding information. A few days later, Pienciak was removed, without explanation, from his post. The investigative team was disbanded, and Gonzalez was left to cover the story on his own. His column was moved to the back pages of the newspaper. Gonzalez laments that the rest of the New York press—*New York Times*, the *Wall Street Journal*, the *New York Post*, and *Newsday*—failed to cover the story. He holds the lack of media coverage responsible for helping to maintain EPA's deception (Gonzalez, 2002: 18–19, 23).

Ironically, in 2007, the *Daily News* was awarded a Pulitzer Prize for its coverage of the health hazards surrounding the events of 9/11. In responding to the award, Gonzalez said, "My only concern is that, if more journalists, not just at the *News*, but the rest of the New York Media, had the courage to follow up on the story back then, maybe there wouldn't be as many people getting sick or dying now" (Hagey, April 17, 2007).

In 2004, in a federal district court ruling, Judge Deborah A. Batts allowed a class action suit on behalf of residents and schoolchildren in lower Manhattan to go forward. The suit alleged that Whitman, other EPA officials, and the EPA failed to warn people about the hazards from the fallout from the Twin Towers and failed to perform an adequate cleanup.

Judge Batts wrote, "The allegations in the case against Whitman's reassuring and misleading statements of safety after the September 11, 2001 attacks are without question shocking." Batts went on to proclaim that "by these actions [Mrs. Whitman] increased, and may have in fact created, the danger" to people in lower Manhattan (Preston, February 3, 2006). Batts was not alone in her outrage. One woman, a volunteer and emergency responder, to whom I spoke said that she toiled for days at the worksite unaware of the dangers. "Why did they not warn us? Had I known I would have worn protective equipment or not have gone back. I am a single mother, who can no longer work because of my respiratory disabilities. I trusted them when they reassured us and they lied to us" (Hollis, March 15, 2004).

A firefighter, who was retired at the time of 9/11, also volunteered to assist "his fellow firefighters." He spent two weeks sorting through the rumble searching for "fallen comrades." He is now permanently disabled and, because he was a volunteer, not eligible for medical treatment and benefits. "I served two tours of duty in Nam and twenty five years fighting fires and this is how my nation treats me?" (Watson, March 15, 2004).

Ironically, the four federal health officials agreed with Jenkins that the concentration of the asbestos dust found in residences and offices near Ground Zero was comparable to the levels found in Libby, Montana. This conclusion caused critics to question why Libby was declared a Superfund site while, at the same time, the EPA was dismissing the same hazardous levels of asbestos found in lower Manhattan.

Information Withheld in the Wake of Katrina

Four years later, in the wake of Hurricane Katrina, the EPA again withheld key information about the hazards of asbestos from residents of the Gulf Coast. This failure, came, astonishingly, shortly after Congressional hearings criticizing the EPA's handling of air quality issues after 9/11. The Government Accountability Office (GAO) report criticized the Agency, this time for inadequate air quality control monitoring around demolition sites along the Gulf Coast. EPA was criticized for not disseminating information about toxic contamination in an adequate or timely manner. The EPA seems to have fulfilled the mission assigned to it under the National Response Plan, created after 9/11 to respond to national crises effectively, by monitoring the air and hazardous substances, in the early hours and days of Hurricane Katrina coming ashore. However, the GAO report stated that the EPA failed to continue such monitoring throughout the recovery period, during which the Agency was supposed to monitor closely for toxic contamination in and around demolition and renovation sites in New Orleans neighborhoods. While apparently it began such monitoring in the months following the storm, it gradually decreased its efforts by the summer of 2006. The GAO accused the EPA of not "aggressively" monitoring for asbestos. Moreover, GAO stated that while the EPA made some effort to inform residents and workers of health risks, it did not adequately fulfill this obligation. According to the GAO, the first environmental assessment took three months to complete and was criticized for confusing and contradictory information, not unlike the information that bewildered St. Bernard residents in the wake of the Murphy Oil spill. The GAO's criticism reflected the same criticism that was made of the EPA in the wake of 9/11, mainly that the Agency played down the risks and created a false sense of security. The EPA's troubles would not end there. Not only did it come under intense criticism for withholding information about toxic trailers in the wake of Hurricane Katrina, but it would also be confronted with severe criticism for withholding information about dispersants during the BP Deepwater oil spill.

While the EPA reports stated the "majority" of exposed sediment tested was safe, it later (eight months later!) stated that such measures of safety referred to short-term visits, not to exposure for those individuals living in or near the contaminated area. Somewhat of a similar disclaimer was employed by the EPA after 9/11 when the Agency qualified its statements about exposure levels by stipulating that they were only for short-term exposure—a condition hard to imagine for the clean-up workers, nearby residents, Wall Street employees, and schools in the vicinity. The 2005 EPA assessment also was flawed because in its examination of buildings, it collected flood sediment data from outside as well as inside the buildings, which is highly questionable since it is well known that buildings act as contaminant traps. Such an approach distorts the higher levels of contamination inside the buildings and makes the overall contamination levels seem lower.

The GAO also criticized the EPA for not having air quality monitors in areas "undergoing substantial demolition and renovation" (U.S. GAO, 2007:2). Ambient air quality monitors are only capable of effectively detecting asbestos if they are located close to demolition sites. The GAO also took issue with the fact that thousands of homes that were being renovated by or for homeowners were not subject to EPA emission standards. Therefore, monitors were not placed nearby. Finally, as you may recall from earlier chapters, under orders from President Bush, the EPA and OSHA suspended air quality laws, which, in part, allowed for faster demolition and building without the normal requirements for asbestos testing and removal. Residents and workers were, thereby, made more vulnerable to toxic hazards. Once again, the EPA was found to provide over-generalized and misleading information about potentially fatal toxic substances to the public. The EPA's troubles would not end there. Later, in 2010, the Agency would fall under severe criticism for not fully disclosing information about the health hazards of dispersants used in the BP Deepwater spill cleanup.

The Mount Sinai School of Medicine issued a report that stated, two-and-a-half years after 9/11, that 70 percent of more than ten thousand World Trade Center responders reported new or worse conditions. The study did not include New York City firemen who were monitored in an independent study by the fire department. Nearly 30 percent of the nonsmokers in that group were diagnosed with breathing difficulties; 30 percent of the patients in the study demonstrated diminished lung capacity (Herbert, et al., 2006).

Devra Davis in her account of the deadly Donora, Pennsylvania, smog (1948) writes that twenty people died in the immediate aftermath and fifty more died a month later. She speculates thousands may have died over the decade that followed. The official public health records only record the deaths of the first twenty individuals to die. The records, and unfortunately history, have ignored the deaths of the others (Davis, 2002: 29). One wonders how many individual deaths in the years after 9/11 will be remembered and how many, if any, will be recognized in the history of that terrible tragedy. Most likely, they will be lost in the unofficial history of the event.

Government Agencies and Responsibility

As has been demonstrated, government agencies are not value-free, neutral entities. They are highly susceptible to political influence from within and without. Congressional investigations, hearings, and independent investigations have also been highly critical of the ATSDR for its continued failure to protect the public from toxic contamination. In 1980, when Congress passed the Superfund law, Congress assigned the EPA with cleaning up the majority of the sites. Simultaneously, Congress also created within the Public Health Service the Agency for Toxic Substances and Disease Registry (ATSDR). For the first three years, ATSDR did not materialize. Dr. Vernon L. Houk, a CDC employee who was in charge of setting up the agency, did not feel there was a need for such an agency because he was of the strong opinion that chemicals were of little harm to public health.

In 1983, the Environmental Defense Fund, the Chemical Manufacturers Association, and The American Home Petroleum Institute filed a lawsuit seeking compliance with Congressional mandates. Houk reluctantly established the agency but provided little leadership and guidance. His reluctance drew criticism from the National Academy of Sciences and public health officials and scientists across the nation. Exasperated, Congress assigned the agency additional responsibility and provided explicit requirements: a) conduct health assessments of every site on, or proposed, for the National Priorities List; b) establish a priority list of chemicals at Superfund sites; c) for each substance on the list, provide a toxicological profile; and d) undertake studies of the health effects of both hazardous waste sites and hazardous substances (Environmental Research Foundation, 1992: 2).

In many ways, the assignment was unrealistic, especially given the time frame within which Congress demanded the agency to respond. For instance, The ATSDR was only given two years to produce health assessments of all 951 sites that were listed as of October 1986. It was given only one year to do the same for any site added after that date. The latter requirement was unrealistic because it required ATSDR to provide health effect studies at a site before the EPA could complete even its initial evaluation of that site. Moreover, if the ATSDR was to meet its deadline of creating health assessments for all 951 sites, it would literally be required to produce two health assessments every day! Ignoring its own guidance requiring site visits, the ATSDR substituted desk assessments. Making matters worse, no internal quality checks or outside expert reviews were conducted as well (Environmental Research Foundation, 1992: 3). Some critics speculate that such impossible tasks ensured that the ATSDR would fail in its mission and were the primary motivation for private industrial associations joining the Environmental Defense Fund suit against ATSDR.

The Congressional mandate was a guaranteed prescription for either failure or superficial results at best, which is exactly what happened. Consequently, ATSDR employed techniques that were extremely inadequate to the task. However, this early history did achieve some desired results for two of the original petitioners, the Chemical Manufacturers

Association and the American Petroleum Institute. Simply put, Congress's response, as a later critical report stated, absolved the EPA from being required to create public health assessments while conducting their site assessments (Russell, et al., 1992), which in turn led to some risk assessments that provided meaningless data.

The second major failure was, in the words of some experts, the creation of "a massive scientific fraud." From the 951 sites that ATSDR studied, the data for evaluating environmental risk and deleterious harm to communities was adequate for only about a third of the sites. The majority of the remaining sites turned out to be inadequate "to support ANY conclusions" (Environmental Research Foundation, 1992). The outcome of most of the studies was, therefore, highly flawed. The Agency's conclusion that the majority of sites presented no public health threat was unsubstantiated.

This failure drew harsh criticism from highly respected public health researchers. The most damning assessment came from an investigative study written by the Environmental Health Network and the National Toxics Campaign (Russell, et al., 1992). This extensive, well-documented report accuses the CDC and ATSDR of "systematically engaging" in practices that were "inconclusive *by design*" (emphasis original, 1992:1). The bold statement in the report's abstract conveys the harshness of the charges: "These intentionally inconclusive studies have been used by polluters and government officials to mislead citizens into believing that further measures to prevent toxic exposure are unnecessary" (i), a charge whose undercurrents still haunt the ATSDR. The report begins by alleging that the agencies "engaged in politically driven whitewashes rather than systematically applying public health principles consistent with their legal and ethical duties" (Russell, et al., 1992: vi). In support of this allegation, the report attributes several factors to their flawed study design:

a) inadequate contact with populations being studied; b) reliance on testing techniques entirely inappropriate to the type of exposure that is involved; c) reliance on statistical methods of inquiry which are entirely unsuited to the small and mobile populations residing around waste sites; d) contracting with researchers who are known to be biased against finding any connection between toxic pollution and disease; e) studying the wrong types of illnesses, e.g. focusing on death studies where health problems experienced to date have been non-lethal, such as respiratory disease. (Russell, et al., 1992: vi)

The report continues by first taking the CDC to task for what it terms "whitewashes of pollution impacts." It cites some of the classic cases of environmental pollution at that time, cases that remain to this day superlative examples of communities afflicted by toxic waste. For example, "Inconclusive by Design" reviews the dioxin contamination at Times Beach, Missouri, where citizens were evacuated because of the concern for their exposure to dioxin, one of the most deadly toxins known. According to the authors, the CDC concluded that the residents of Times Beach suffered no ill health effects as a result of their exposure. However, other studies conducted demonstrated evidence of immune system abnormalities. The report faults the CDC for cutting residents from the study because they were too ill to participate, including four individuals who had symptoms of a skin disease that is caused by dioxin!

The report also summarized the CDC's health studies in what has become the classic case study of a contaminated community: Love Canal. The report alleges, among other things, that the CDC "attempted to turn a physical study" into a mental health study (Russell, et al., 1992: vii). Yet another classic case was also examined: Agent Orange, which Congress had directed the CDC to investigate for its potential health effects on Vietnam Veterans. Agent Orange contains a significant amount of dioxin. In this instance, the Agency's harshest critic was Elmo R. Zumwalt, Jr., admiral in charge of Naval Operations during the Vietnam War, who called the investigation "a fraud." He further alleged that Vernon Houk, who was in charge of the study, "made it his mission to manipulate and prevent true facts from being determined" (Russell, et al., 1992). Furthermore, as cited in the report, Zumwalt testified before Congress that Houk either withheld in some cases or manipulated in other cases, the data surrounding the facts in order to contend that a large-scale study of the problem was impossible (Russell, et al., 1992: viii).

The independent report cited above is not the only report to have criticized the ATSDR. An earlier report in 1991, issued by the GAO, concluded among many other things that the Agency's assessment was weak, incomplete, and of limited value. Another report issued by the National Research Council of the National Academy of Sciences (1991) criticized the CDC and ATSDR for failing to identify, assess, or rank hazardous waste sites and their potentially harmful impact on communities properly. Congress charged the Academy to review and evaluate the data for an estimation of the potential health impacts at Superfund sites. The Academy's report demanded a more "prudent" public health policy in order to assure the health of Americans. The report concluded that, contrary to the findings of the CDC and the ATSDR, hazardous waste sites were producing harmful effects on U.S. citizens and that the agencies' approach to the Superfund sites may be leaving many citizens vulnerable to potential adverse health effects.

Among the many other criticisms that "Inconclusive by Design" makes is in regard to the ATSDR's reliance on state health departments to conduct the vast majority of its studies. In the late 1980s, the Agency created a cooperative agreement with a number of states to conduct health assessments and investigations. This decision has come under fire from a number of quarters including from public health professionals. As the report states, state public health agencies are a poor choice for conducting such investigations and assessments for a number of reasons. Foremost among these is the fact that, while the decision to rely on public health departments was justified in part because state agencies were more directly accountable to the public, the opposite may, in many cases, be true. State public health departments are vulnerable to political influence from governor's offices and the state legislature. In the case of the latter, it is imperative to keep in mind that state legislators are more vulnerable to lobbyists and special interests than federal elected representatives. This influence is especially significant in states where special interests, like the energy and chemical industries, are often a dominant player in state economies. "Inconclusive by Design" asserts that evidence demonstrates that state officials are not always accountable to communities as much as they are accountable to polluters (Russell, et al., 1992). These same charges have been made by critics in regard to the case of the health investigations being conducted on the TVA ash spill and in other states where the coal and energy industries are dominant players.

A number of similar criticisms of ATSDR are still being made almost twenty years after the original allegations.. Hearings were held in March 2009 by the U.S. Subcommittee on Investigations and Oversight of the Committee on Science and Technology. The investigation and report are an offspring of the earlier investigation the Committee made on the role of the CDC and its sister agency, ATSDR, in handling the controversy surrounding formaldehyde and the FEMA trailers used in the wake of Hurricanes Katrina and Rita. The report issued by the Majority Staff of the Committee echoed many of the same complaints. Its introduction stated that local community groups across the nation believe that ATSDR has failed to protect them from toxic hazards. The Committee further contended that independent scientists "are often aghast at the lack of scientific rigor in its health consultations and assessments." Similar to criticism made decades before, the report stated that ATSDR's studies lack the ability to make a connection between toxic exposure and illness and that, furthermore, its methodologies are, in the words of the report, "doomed from the start" (U.S. Congress, House, March 10, 2009:1). These criticisms seem to have been either largely ignored or incompletely resolved.

The report reminds ATSDR and the panelists that the Agency's stated mission "is to serve the public by using the best science, taking responsive public health actions, and providing trusted health information to prevent harmful exposures and disease related to toxic substances"(ATSDR Statement of Mission, undated). At the beginning of the report, the committee made it clear that it is not suggesting that the Agency find problems where none exist or that it should, or could, identify a disease or health hazard in every case investigated or in situations where exposure evidence is ambiguous (U.S. Congress, House, March 10, 2009: 2). That said, the report stated unequivocally:

> Yet time and time again ATSDR appears to avoid clearly and directly from confronting the most obvious toxic culprits that harm the health of local communities throughout the nation. Instead, they deny, delay, minimize, trivialize or ignore legitimate health concerns and health considerations of local communities and medical professionals.

The report goes on to reiterate what earlier critics alleged: that independent scientists, public health advocates, medical professional, and community environmental groups believe that "ATSDR often seeks ways to avoid linking local health problems to specific sources of hazardous chemicals." The allegations continue by citing a "current" ATSDR official who anonymously testified to the committee the following: "It seems like the goal is to disprove the communities' concerns rather than actually trying to prove exposures." Apparently, old problems continue to haunt ATSDR and undermine its credibility.

The Committee's report asserts that the reviews of the FEMA trailer scandal, as well as deficiencies in other ATSDR health reports, "suggest" that the Agency has not fully recovered from its earlier problems. Moreover, the report continues by reviewing present-day deficiencies outlined by current ATSDR professionals as well as by outside critics. Employees testified that ATSDR lacks appropriate controls, conducts inadequate analysis of community health risks, and often fails to collect and analyze the "most relevant and revealing data" concerning potential environmental hazards. Local community members testified that ATSDR appears to "do more harm than good by offering them reassuring but unfounded and unsound advice and analysis which creates a false sense of certainty and safety to communities that is not grounded by "scientific inquiry or independent examination" (U.S. Congress, House, March 10, 2009: 4). The report's conclusion based on extensive testimony is extremely damning:

> The longer the ATSDR continues to pursue its role in protecting the public health as it has for the past three decades, issuing deeply flawed scientific reports, not responding to the concerns of local communities and approaching potential environmental exposures with a mindset that endeavors to disprove any link between the public's ill health effects and potential exposure to environmental contaminants or toxins, the more people will suffer. ... ATSDR seems to represent a clear and present danger to the public's health rather than a strong advocate and sound scientific body that endeavors to protect. (U.S. Congress, House, March 10, 2009: 28)

11. A Gulf of Uncertainty
"New Oil in Old Barrels"

Many of the facts and uncertainties surrounding the BP Deepwater catastrophe dramatically underscore several of the recurrent themes in this book. The Deepwater Horizon offshore rig blowout in April 2010 killed eleven men and soon became the largest oil spill in U.S. history. At the time of this writing, it has been spewing oil for over two-and-a-half months and has no doubt exceeded in size the Ixtoc I (1979) blowout in the Gulf of Mexico. The actual size of the spill still remains a mystery, which typifies much of the uncertainty surrounding the event. By now it is reasonable to assume that the Deepwater rig has released over 200 million gallons of crude oil. BP is reported to have used two million gallons of dispersants, the most ever used in an oil spill. Despite an EPA directive issued on May 26 stating dispersants should be rarely used, the Coast Guard approved sixty-four exemptions in forty-eight days, allowing BP to disperse hundreds of thousands of gallons of the chemical on the Gulf's surface. These exceptions include occasions on which BP applied to the Coast Guard for permission after it had already used the dispersants (Wald, August 1, 2010). As of this writing, how this toxic brew has impacted, or may ultimately impact, the environment and coastal populations remains uncertain.

The tragedy struck a particularly responsive chord in the minds of many: Hurricane Katrina increased the symbolic value of the Gulf of Mexico and helped to amplify its vulnerability to future disasters. The Gulf is also home to millions of migratory sea turtles and other marine species as well as one of the nation's richest commercial and recreational fisheries (U.S. Senate, Testimony of Stanley Senner, July 27, 2010).

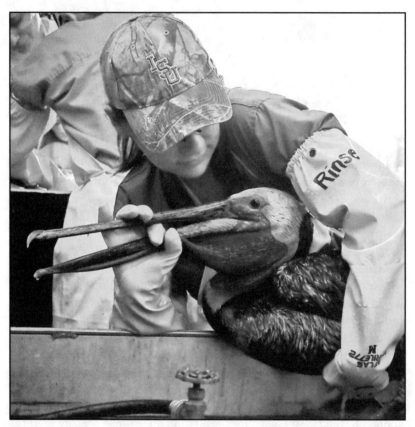

Pelican rescued from the BP Deepwater Oil Spill.
Photograph © Megan Richardson.

The figures surrounding the Deepwater blowout and oil spill are incandescent. BP's COO, Doug Shuttles, made the standard claim that there had been "few big spills over the last few decades" (Maddow Show). Tony Hayward, BP's CEO at the time of the spill, argued that the spill was "unprecedented." Even (now retired) Admiral Thad Allen of the U.S. Coast Guard, who is in charge of overseeing the cleanup effort, has struck a similar note, stating the spill is both "unprecedented and anomalous." These statements ignore the fact that spills, like disasters, are a normal, if unacceptable, part of everyday life. Oil spills are not isolated events, and thousands of spills have occurred in the last four decades alone. These denials also ignore the fact that in addition to permissive regulations, human negligence, and technological failures, offshore drilling is vulnerable to natural forces, which also undermine their safety.

Perhaps the best example is the damage inflicted by hurricanes Katrina and Rita in 2005 when 167 offshore platforms and 450 pipelines were destroyed or damaged, resulting in nine major spills and the release of seven million gallons of oil into the Gulf. Even more dramatic, the storm carried one offshore rig sixty-five miles before dumping it on Dauphin Island. One of Chevron's deepwater platforms, which was operating in 2,100 feet of water, was severed from its moorings, capsized, and drifted almost eighty miles. The year before, Hurricane Ivan damaged twenty-four structures and destroyed seven offshore platforms (Juhasz, 2008: 316–17).

Minerals Management Service (MMS) records also contradict the assertion of the oil industry that offshore drilling in the Gulf of Mexico has a history of safe, clean production. Just a few examples belie this claim. In 1967, a Humble offshore corroded pipeline leaked 160,638 barrels in thirteen days. A Union Oil well blowout spilled 80,000 barrels in 1969 and leaked oil for four years, killing 4,000 birds. A Penzoil pipeline in 1974 spilled 19,833 barrels, and an Amoco pipeline break spilled 15,576 barrels in 1988. In 1990, a Shell pipeline spilled 14,423 barrels and created an oil slick twenty-five by fifteen miles. The accuracy of MMS statistics in reporting spills in the Gulf was called into question on at least one occasion—in 1970 a Shell Oil blowout killed four people and was reported by MMS to have spilled 53,000 barrels of oil. However, Robert Bea, a University of California Berkeley professor who tracked the spill for Shell at the time, disputes that figure and claims the spill was ten times that size (Mufson, July 24, 2010).

One of the most dramatic examples of the frequency of oil spills is the months before and after the Union Oil spill in Santa Barbara in

1969, the first offshore spill in the modern era. While the number and magnitude of oil spills in any given period varies greatly, the average frequency of oil spills is far greater than the media report and the oil industry would have us believe. In March 1967, approximately one year before the Santa Barbara spill, the supertanker the *Torrey Canyon*, also owned by Union Oil, went aground off the southwest coast of England, spilling thirty-five million gallons of crude oil and creating what was then the world's worst oil spill. Less than a year later, off the coast of Nova Scotia, the Liberian tanker *Arrow* went aground, spilling six million tons of crude into the north Atlantic. A week later, a Chevron drilling platform off the coast of Louisiana caught fire and burned for seven weeks, spilling 30,000 gallons of oil.

Next, a Greek tanker spilled oil in Tampa Bay, Florida. Exactly a week later, a Mobil Oil Company drilling pier fire, just eighteen miles from Santa Barbara, caused over a million dollars in damage. Two weeks later, an oil slick was discovered in Alaskan waters, polluting over 200 miles of Kodiak Island. Then, on May 29 of the same year, an offshore explosion and oil spill near Galveston, Texas, killed nine people and injured more than a dozen. The Galveston spill was followed in September of 1970 by a fire and subsequent spill from an oil platform just a few miles from the Santa Barbara spill. Then the unthinkable happened in early October: the collision of two oil tankers in the English channel spilled 70,000 tons of crude into the channel (Easton, 1970: 207–8). The year ended with a blowout, fire, and spill on a Shell platform thirty miles off the coast of New Orleans, Louisiana. It took almost a month to put the fire out. Eight of the rigs' twelve wells caused what was at the time the largest oil spill in the Gulf of Mexico (Dye, 1971: 111). Altogether, the USGS reported that in 1969 there were 1,007 oil spills in U.S. waters. In 1970, there were four times as many (Easton, 1972: 208).

In 1969, the second report of President Nixon's panel on oil spills issued the following sober warning:

> Since 1954, approximately eight thousand offshore oil wells have been drilled. Twenty-five blowouts have occurred, of which seventeen leaked gas only. Two resulted in serious oil pollution incidents and nine constituted serious blowouts that persisted for days with fires (nine cases) or fire hazards and hazards to personnel (twenty-nine deaths). If offshore development continues to expand at the present rate and the frequency of accidents remains the same, three thousand

to five thousand wells will be drilled annually by 1980 and we can expect to have a major pollution incident every year (Dye, 1971:32).

In reviewing the advantages and disadvantages of allowing future development of oil resources, the report states, "postponing some of our development might serve best the future of our country (Dye, 1971: 32)." Another report written the year before the Santa Barbara oil spill by the Department of the Interior (DOI) Secretary Stewart Udall made the prophetic statement that still resonates in our present day predicament: "This country is not fully prepared to deal effectively with spills of oil or other hazardous materials—large or small—and much less with a *Torrey Canyon*-type disaster. Because sizeable spills are not uncommon, and major spills are an ever present danger, effective steps must be taken to reduce this nation's vulnerability (Cowan, 1972: 220)." Another report, "The Economy, Energy, and the Environment," prepared in 1970 by the Joint Economic Committee of Congress, warned that the nation should adopt a "go slow" policy in the development of offshore rigs (Dye, 1971: 1973).

The number and severity of oil spills may vary annually, but records indicate that the amount of oil spilled per year has not decreased in recent decades, contrary to industry claims. Between 1978 and 1994 alone, there were thirty-four incidents in which over ten million gallons of oil were spilled (Burger, 1997: 29). In terms of offshore oil rig spills, there has been a marked increase within U.S. waters. A recent *USA Today* report states that during the last decade, spills from offshore rigs have more than quadrupled (Levin, June 8, 2010). *USA Today* based its report on an analysis of federal data, which also shows that oil spills grew worse even when taking into account the increased production of offshore oil drilling. The article also reports that the increase in spillage should have served as a warning for the Deepwater leak. BP, the oil company with the most spills in the period 2000–2009, has the worst safety record of any major oil company (Dickinson, June 8, 2010).

Clearly, oil tanker and offshore rig spills are common enough to be of great concern and warrant stringent regulations. The Montera offshore rig spill in the Timor Sea, off the coast of Western Australia, spewed out of control for seventy-four days beginning in August 2009. It occurred only eight months prior to the Deepwater tragedy. Although the spill did not occur in U.S. waters, it was a precedent to the Deepwater spill. Some apologists argue that the number of tanker spills or offshore blowouts that have occurred in U.S. waters alone are statistically insignificant. Nothing could be further from the truth. Moreover, the point is not how many spills occur in U.S. waters, but that the oil companies and players at fault are often the same worldwide, as are the faulty technologies responsible for the spills. To compartmentalize the world's oceans apart from the historical legacy of oil spills is an unrealistic basis for preventative policy.

Finally, the media and oil industry seem to have forgotten two of the largest offshore oil platform disasters in history. In 1980, the Norwegian drilling rig Alexander L. Kielland capsized, killing 123 people (Carson, 1982: 5). An even worse disaster occurred in 1988, when the Piper Alpha oil rig in the North Sea, operated by Occidental Petroleum, exploded; although the accident spilled no oil, the explosion and ensuing fire destroyed the rig and killed 167 men (Appleton, 1994).

Downplaying Disasters

As we saw in the accounts of the *Exxon-Valdez* spill, the Shetland Islands spill, the Murphy Oil spill, the controversy over the FEMA trailers, and the TVA spill, as well as the health controversies following 9/11, those parties responsible for a disaster, or for the response to a disaster, commonly downplay the events even though such denials seldom seem to succeed. BP's attempts to do likewise were not the first attempt by an oil company to downplay the impact of an offshore blowout.

In 1969 in the wake of the Santa Barbara oil spill, Fred Hartley, the president of Union Oil, stated that he was "impressed" with all the publicity the dead birds had received in the wake of the spill when no humans had perished. Hartley found it significant that even though there were no human fatalities people perceived the event as a disaster: "Although it has been referred to as a disaster, it is not a disaster to people. There is no one being killed" (Potter, 1973: 299). He thereby dismisses the social, psychological, and economic impact of the spill. The eventual fallout from the spill proved Hartley wrong.

BP's attempts are in keeping with this well-established pattern of trying to minimize the impact of a disaster. Tony Hayward, in one of many gaffes, stated, "The Gulf of Mexico is a very big ocean. The amount of volume of oil and dispersant we are putting into it is tiny in relation to the total water volume" (Webb, May 14, 2010). This statement was followed only days later by the following: "[T]he environmental impact of this disaster is likely to be very modest" (Milam, May 18, 2010), a statement that both he and BP came later to regret.

One might view such statements as typical of the usual litany of PR we have come to expect in the wake of disasters and an attempt, no doubt, to create an on-record response for anticipated lawsuits. Nonetheless, some commentators have their own motivations for downplaying disasters. Consider, for example, the statement of commentator Rush Limbaugh: "This spill is nothing more than an opportunity for the Left to continue to attack this country and the people who make it work." Rand Paul's statement on "Good Morning America" expressed a similar cynicism: "I think it's part of this sort of blame-game society in the sense that it's always got to be somebody's fault instead of the fact that maybe sometimes accidents happen" (Stein, May 21, 2010).

Mississippi Governor Haley Barbour's initial response to the Deepwater blowout was that it was overblown. He blamed the national press corps for exaggerating the impact of the spill: "The biggest negative impact for us is the news coverage" (Stein, June 6, 2010). Governor Barbour changed his mind when the spill has come ashore in his state. He is now complaining about the failure of the federal government's response (Hennessy-Fiske and Fausset, June 28, 2010).

Inadequate Contingency Plans

Despite the American Petroleum Institute's assertion that "companies operating in these federal waters must comply with a rigorous set of preparedness and planning requirements," this proves not to be the case (American Petroleum Institute, 2009). The failure to prepare adequately for a disaster, or as Lee Clarke observes, "the failure to imagine the unimaginable," is lamentably a common theme in most disasters (2006: 22). This point is well illustrated in the case of the Deepwater disaster. Rick Steiner, a veteran of the *Exxon-Valdez* spill and a retired marine science professor, commented on BP's oil spill contingency plan: "This response is not worth the paper it is printed on. ... Incredibly, this voluminous document never once discusses how to stop a deep water blowout even though BP has significant deep water operations in the Gulf" (*Huffington Post*, May 25, 2010).

The House Energy Committee hearing investigating BP's emergency plans revealed not only BP's failure but the failure of just about the entire industry to prepare adequately for a worst-case scenario. In a moment of unusual candor, Rex Tillerson, the chairman and CEO of the world's largest oil company, Exxon-Mobil, under questioning from the committee admitted that the industry is ill equipped to contain the damage from a major spill (Tilove, June 16, 2010). BP CEO Hayward, in an understatement to the press, said, "What is undoubtedly true is that we did not have the tools you would want in your toolkit" (Wearden, June 3, 2010). In fact, in BP's 582-page plan for responding to a spill in the Gulf of Mexico, the section on "worst case" scenarios was left blank!

Sociologist Harvey Molotch's analysis of the Santa Barbara offshore rig blowout four decades ago rings true today. He pointed out then the striking contrast between the cleanup methods used and the technology employed to find and drill oil (Molotch, 1972: 135): "[T]he sophistication of the means used to locate and extract oil compared to the primitiveness of the means to control and clean it up was widely noted in Santa Barbara. It is the result of a system which promotes research and development which leads to strategic profitability rather than to social utility."

The House hearings dramatically revealed for the first time just how meaningless the oil spill contingency plans of all the leading oil companies actually are. They might be characterized best as "cut-and-paste" documents that adhered to a "one size fits all" approach, regardless of the region or conditions to which they apply. As Rep. Edward Markey (D-Mass) quipped, "The only technology you seem to be relying on is a Xerox machine" (Tilove, June 16, 2010). Absurdly, the Gulf plan included information on to how to protect walruses, a reference obviously lifted from plans intended for coastal Alaska. More concerned with PR spin than with realistic, meaningful responses, Exxon-Mobil's forty-page appendix described how to deal with reporters in the event of an accident; only five pages were devoted to protecting resources, and nine pages to oil removal (Dlouhy, June 15, 2010).

Clearly, in most of the disasters discussed earlier in this text—especially the *Exxon-Valdez* and TVA ash spills—the emergency response plans proved inadequate to the challenges that confronted the responders. Perhaps "the failure to imagine the unimaginable" is simply wishful thinking on the part of corporate managers. It is possible they place too much faith in their engineers' creative know-how. Or, perhaps, as some cynics have stated, they believe that it is cheaper to pay the cost of the unimaginable, should it occur, than invest in the research and development that could either prevent or remedy a disaster.

Lee Clarke, whose book *Worst Cases* should be mandatory reading for all corporate and government officials, offers considerable insight into our culture's failure to prepare for worst case events. One of his many trenchant observations is that accurate worst-case thinking is different from our current approach to risk. Our modern approach is based on probabilistic thinking—that is, what the likelihood of something happening will be. On the other hand, worst case thinking needs to be based on thinking in terms of possibilities, or what happens if something really goes wrong whether or not it is likely; this happens, Clarke wisely counsels, more often that we seem to want to admit. In other words, disasters that we think are improbable actually are not exceptional events but routinely occur.

Some MMS employees were in fact aware of the agency's failure to think in terms of worst case scenarios. In May 2000 an environmental assessment study recognized the real potential of a spill of a very large magnitude. Even the oil industry estimated that a spill could range anywhere from 5,000 to 116,000 barrels a day and could result in undersea plumes. The assessment was realistic enough to state, "[t]here are few practical spill response options for dealing with submerged oil" (Dickinson, June 8, 2010).

During this period, MMS researchers and members of the oil industry—including the company then known as BP-Amoco—acknowledged that a large spill could threaten the future of offshore oil drilling. The document stated, shockingly enough, that the oil industry "could ill afford a deepwater blowout" and recognized quite candidly that "no single company has the solution" to such a situation. It further stated that a deepwater blowout would present the ultimate test (Dickinson, June 8, 2010).

According to Dickinson's account, the Bush administration ignored these early warnings. In this climate of denial, the MMS under Bush dinted a large oil spill as around 1,500 barrels a day in the worst case scenario. Moreover, in 2007, MMS officials went so far as to say that "blowouts were a low probability and a low risk." Furthermore, the agency's environmental assessment for BP's lease block stated that offshore spills "are not expected to damage significantly any of the wetlands along the Gulf Coast" (Dickinson, June 8, 2010).

If these projections seem too naïve or ignorant, consider this: President Obama, after his appointment of Secretary of the Interior Ken Salazar, reassured the public by saying, "It turns out, by the way, that oil rigs today generally don't cause spills. They are technologically very advanced" (Dickinson, June 8, 2010). This statement totally ignores the facts, many government documents, and government sponsored studies. One must question the credibility of the president's top advisors.

Access to Information

As we have witnessed, gaining access to information in the wake of disasters has long been a major contributor to uncertainty. When the afflicted populations, the media, the public, and sometimes even government agencies have difficulty obtaining accurate information, responding effectively to the challenges presented by disaster is difficult. Moreover, the withholding or manipulation of information allows certain parties, especially those who are to blame or those responsible for remediating the situation, to create an official narrative that is difficult to contest. The media's ability to access information has been problematic in many of the disasters discussed in this book. In the case of the Shetland Islands oil spill, the U.K. government's withholding of information, particularly in regard to health issues, coupled with the lack of a freedom of information act, made gaining access to information difficult for both journalists and the public. The withholding of information by both FEMA and ATSDR proved troubling in the controversy surrounding the use of FEMA trailers. Similar withholding, or nondisclosure, of information about environmental health issues in the aftermath of 9/11 proved to be equally troubling. In the case of the chemical explosion in Bhopal, Union Carbide refused to release information about the chemical. The more recent TVA ash spill demonstrates yet again a lack of transparency. Unfortunately, difficulties obtaining information are now an integral part of the disaster unfolding in the Gulf of Mexico.

The withholding of information and outright deception appear to be part of BP's media strategy. What makes the situation exceptionally troubling is that NOAA, which is leading the research of undersea plumes, has refused to share its data with independent scientists and the public. These plumes are of vital concern because scientists suspect that they will devastate deepwater marine life and harm marine animals and the coastline of the Gulf for many years to come. Data obtained in a timely fashion could prevent harm in both the short and long term. Scientists find NOAA's failure to release the data especially disturbing because the Agency is in fact sharing it with BP. Normally it would not share the data with BP because of potential litigation between the federal government and BP. However, because BP has been made part of the Joint Incident Command, the company is allowed access.

Access to accurate information is also thwarted when media events are staged to deceive the government, the media, and the public—for example the staging of an oil spill response for the sake of President Obama's visit to the Gulf coast is a case in point. This drama bears an uncanny resemblance to Exxon's staging of a cleanup on the outer coast of the Kenai Peninsula simply for the sake of a network nightly news coverup. According to several sources, including Chris Roberts, a Jefferson Parish councilman, BP bused in hundreds of temporary cleanup workers immediately prior to the President's visit to Grand Isle, Louisiana, at the end of May (Linkins, May 28, 2010), and removed them from the scene shortly after his departure.

Such theatrics seem minor compared to numerous reports from a variety of media outlets that BP, with the assistance of public officials, has been trying to restrain access to spill sites. The earliest reports of these tactics came in late May when journalists from *Mother Jones Magazine* reported being "stymied" from gaining access (McClelland, May 24, 2010). According to reporter Marc McClelland's account, Jefferson Parish sheriff's deputies informed him that he could not go to Elmer's Island to view the spill. Approaching the island, he encountered a four-car police blockade. Police informed him that he had to go to the BP Information Center to obtain permission. There, a BP public relations employee informed him that Elmer's Island was closed and could be visited only with the escort of a BP representative. When he asked why he needed BP's permission—since it was not BP's land—he was informed that BP was in charge "because it is BP's oil" (McClelland, May 24, 2010).

In another incident, BP contractors and Coast Guard officers told a CBS film crew attempting to film a heavily oiled beach that they had to leave. This incident, like the one above, brought into question the relationship between BP, the armed forces, and local police. Rob Wyman, the lieutenant commander of the U.S. Coast Guard Deepwater Command Joint Information Center, responded to the CBS incident by saying that the Coast Guard did have rules that prohibit journalists from access to heavily impacted places. Ostensibly, such rules were enacted for safety reasons. He further explained that many hundreds of journalists to date had been "embedded" and allowed to cover response efforts. Of course, the irony here is that many tourists and local people were allowed access to areas that were restricted for journalists, undermining the safety explanation.

The notion of embeddedness disturbed many in the media because in prior disasters, the media was not required to be embedded. To many journalists, the notion of embeddedness invokes the constraints usually imposed on media coverage of military engagements. Being embedded means that journalists are often allowed only to view the news from an official point of view. In this case, the perspective was not the military's, but that of a private corporation. The practice also raises the question about whether someone has to have a press pass in order to be embedded, thereby theoretically restricting the public's access to documenting a disaster.

The Louisiana American Civil Liberties Union, concerned about BP's attempt to censor, released the following statement:

> In the United States, we value free access to information and we rely on an uncensored media to provide a full picture of matters of public importance. Answers are provided through more information, not less. The public has the right to know what's happening on the Louisiana coast. BP cannot impose its own rule of law on the people of Louisiana or the Gulf Coast, just because it doesn't want us to know what is going on. (ACLU, June 28, 2010)

Riki Ott, a marine toxicologist who was involved in the *Exxon-Valdez* oil spill response, accuses "BP of using federal agencies to shield itself from public accountability" (Ott, June 11, 2010). According to Ott, when she was flying in a small plane over the spill region, the pilot noted that the red line marking federal flight restrictions of 3,000 feet over the oiled region had been adjusted to include coastal barrier islands along Alabama's coast. Ott claims the pilot said the flight restriction changes were enacted because "BP doesn't want the media taking pictures of oil on the beaches."

A *Times-Picayune* photographer encountered a similar problem in late May while on a charter flight over the Gulf. Approaching Grand Isle, the pilot told him that the Federal Aviation Administration had imposed flight restrictions preventing fly-overs below 3,000 feet (Kirkham, July 2, 2010). Such altitude restrictions make it impossible to photograph the affected areas.

PBS "Newshour" journalists trying to cover the health impacts of the spill in Plaquemines Parish, Louisiana, have been unable to gain access to a federal mobile medical unit situated in a BP compound area. According to Bridget Desimone with "Newshour," the health unit was established by the U.S. Department of Human Services and is staffed by medical personnel from the HHS National Disaster Medical System. For two weeks, Desimone and her crew attempted unsuccessfully to gain access to the facility to investigate how many people were being treated and what they were being treated for. Access was repeatedly denied the news team. According to Desimone, Dr. Manny Alvarez of Fox News was also denied access (Desimone, June 30, 2010).

The HHS claims the facility was established to provide basic care to responders and residents who were concerned about the health effects of the spill. The arrangement of the clinic, however, seems highly unusual. People seeking treatment at the center first have to be screened by a private company, Arcadian Ambulance services, hired by BP. This process raises disturbing questions about BP's role in the gate-keeping process and about what access BP might have to patients' medical records— patients who may be making future claims against the corporation. The facility is so shrouded in mystery that obtaining information about what is actually going on inside the clinic is extremely difficult.

In an interview with PBS, Dr. Irwin Redlener, president of the Children's Health Fund and director of Columbia University's National Center for Disaster Preparedness, described the abnormally high security surrounding the facility. Even though he is both a physician and a highly regarded member of the disaster response community, Redlener had to go through an extensive security check in order to gain entrance to the clinic. Once inside, he witnessed the peculiar scenario in which potential patients had to go through the private initial screening facility. Based on his interviews with the HHS personnel, his impression was that they did not have a very clear idea of how many people were trying to get treatment. The HHS team members were under the impression that local residents were not seeking their services. From what they heard from local Parish officials, their services were not considered necessary, impressions that, said Redlener, "fly in the face of everything they were hearing in the parish" (Desimone, June 30, 2010).

Reports of media restrictions to spill areas continue to grow. In the beginning of July, the U.S. Coast Guard imposed new rules under which people are restricted from coming within sixty-five feet of response vessels, booms on the water, or beaches. The Coast Guard issued the following statement to justify this ruling: "The safety zone has been put in place to protect members of the response effort, the installation and maintenance of oil containment boom, the operation of response equipment and protection of the environment by limiting access to and through deployed protective boom" (Deepwater Horizon Response, 2010). Obviously, this makes it much more difficult to record or document the spill's impact. If journalists want to gain greater access, they now have to contact the Coast Guard captain of the Port of New Orleans to obtain permission. Violation of these rules could result in a fine of up to $40,000 and could be classified as a class-D felony.

According to a report in the *Times-Picayune*, Gerald Herbert, an Associated Press photographer, has raised questions about the restrictions and requested a meeting with Coast Guard officials (Kirkham, July 2, 2010). Herbert's statement reflects the attitude of other photojournalists with whom I have spoken: "Often the general guise of 'safety' is used as a blatant excuse to limit the media's access, and it has been done before. It feels as though news reporting is being criminalized under thinly veiled excuses. The total effect of these restrictions is harming the public's right to know." Matthew Hinton, also with the Associated Press, has been photographing the spill area and said that the restrictions make photographing birds along the booms increasingly difficult: "[Y]ou'd have to mount a telescope" to the camera to get a photo (Kirkham, July 2, 2010).

A day after this story was filed, another appeared as part of an ongoing collaboration with "ProPublica" and PBS "FRONTLINE" (Engelberg, July 3, 2010; Desimone, June 30, 2010). Lance Rosenfield, a photographer working on the story for these news organizations, was detained while taking pictures in Texas City, Texas. The incident began when he took a picture of a Texas City sign on a public highway and was detained by a BP security agent and a local police officer who "identified himself as an agent for the Department of Homeland security." After his photographs were reviewed and his birth date, Social Security number, and other information recorded, he was released. No charges were filed. BP spokesperson Michael Marr released the following statement: "BP security followed the industry practice that is required by federal law. The photographer was released with his photographs after those photos were reviewed by a representative of the Joint Terrorism Task Force who determined the photographer's actions did not pose a threat to public safety."

The editor-in-chief of "ProPublica" responded with a statement objecting to the procedure and said that if it was necessary for a law enforcement official to review the unpublished photographs he saw "no reason" why the photographs should be shared with BP representatives (Engelberg, July 3, 2010). Anderson Cooper of "CNN News" said the rule prevents reporters from being "anywhere we need to be" and make it "very easy to hide incompetence or failure" (Linkins, July 6, 2010).

In recent years, I have experienced firsthand similar attempts to restrain my research. While conducting research on the Murphy Oil spill, I was prohibited from taking photographs of the Murphy Oil refinery in St. Bernard Parish by company security officials who threatened me with arrest even though I was on public land. I encountered similar threats while attempting to document the effects of the TVA ash spill. A Roane County sheriff threatened me with arrest if I did not put away my camera and return to my car even after I explained to him that I was conducting research.

An even more disturbing incident occurred while I was researching Hurricane Katrina evacuees in the Houston Astrodome. I had followed the research protocol of notifying in advance the facility officials of my research agenda. Oddly enough, I had to make numerous phone calls to discover exactly who was in charge of the incident command center at the Astrodome. Surprisingly, numerous state, federal, and Red Cross officials did not seem to know who was in charge. After finally contacting an Astrodome official, I informed him that I had a National Science Foundation Quick Response Grant administered by the Hazards Center at the University of Colorado at Boulder and that the grant project was also endorsed by FEMA.

In a decidedly unfriendly conversation, I was told that "not just anyone could come to the Astrodome and conduct research" and that "before I got on a plane or set foot in the Astrodome," I had to gain his permission. The tenor of his voice alone told me that permission was not forthcoming. Disturbed by the conversation, I checked with the grant-making officials and was informed that I had every right to proceed with my research and that I should not stand down. When I arrived at the Astrodome, I left my credentials and research protocol on the incident command table—which at the time happened to be unmanned—and proceeded without further incident.

Once inside the Astrodome, I encountered an official from the National Archives, accompanied by a photographer who was documenting the plight of the evacuees. He recounted to me a similar story in which an Astrodome official attempted to discourage his presence.

I have wondered if other researchers encountered similar difficulties and whether some of them were more intimidated by such procedures and turned away. The use of intimidation tactics in the wake of disaster is relatively new in the United States and has a chilling effect on both journalists and researchers. My only previous experience with such restrictions had been while working in dictatorships in Central America and Asia.

Access to Reliable Research Information

In the aftermath of the BP Deepwater disaster, recurring controversies over access to information surround debates about the actual size of the spill and the existence of deepwater plumes. Since the actual size of the oil spilled ultimately could be a determining factor in the size of the fine imposed by the federal government on BP and the amount of compensatory and punitive fines that might result from litigation, it is obviously in BP's best interest to downplay the actual size of the spill.

The scientific uncertainty, as well as the downplaying or misreporting of spill size and flow, contributes to the inability of affected community members, responders, and scientists to respond effectively to an oil spill. Furthermore, the uncertainty that this situation engenders heightens public anxiety. The credibility of both the federal government and BP is on the line.

Uncertainty and controversy over the size and the direction of the oil spill played a major role in the wake of the *Exxon-Valdez* oil spill, as it did in many other similar disasters. After the massive blowout off the coast of Santa Barbara (1969), many critics were convinced that Union Oil Company's estimates were ten times lower than the actual size of the spill. Among these doubters were the head of the Department of Interior's response team, the regional director of the Federal Water Pollution Control Administration, and independent scientists whose calculations held Union's estimates in question. Sierra Club investigators held similar doubts, all of which contributed to mounting distrust over Union Oil's handling of the disaster. Later the President's Panel on Oil Spills would concur that Union Oil's estimates were far too low (Easton, 1972: 142, 250).

In the case of the *Exxon-Valdez* oil spill, controversy about the actual size of the spill existed almost from the beginning. Rumors abounded in Alaska that the divers who examined the tanker's hull reported that eight of the twelve compartments were breached and that the oil spill had to be larger than the official estimate. Then in 1990, two newspaper accounts reported that the spill size was anywhere from twenty-seven to thirty-eight million gallons of oil (Hennelly, 1990; Spencer, 1990). Of the hundreds of people I have interviewed in Alaska over the last two decades, I seldom encountered anyone who believed the spill only consisted of the *official* eleven million gallons of oil.

According to Riki Ott, the Alaska Department of Environmental Conservation hired an independent investigator to estimate the volume of oil spilled. Not until 1994, in response to a public records request, was the investigator's report made public. The 1989 report stated that the tanker left port with 53.04 million gallons of oil onboard. After the spill, 42.2 million gallons were transferred to three tankers, leading the investigator to conclude that 10.8 million gallons were spilled, a figure that never received independent verification.

Later, another state investigator's report stated that the earlier report had "serious deficiencies" (Bluemink, June 6, 2010). State investigators tracked down the three tankers used to transfer the oil. Although Exxon claimed that the offloaded cargo contained 100 percent oil, state investigators discovered there was so much seawater in all the tankers that the refinery refused to offload it. The cargo then remained on the tankers and was eventually returned to the Port of Valdez to be used as ballast water. The document states that the state's conservative estimate was that thirty million gallons of oil were spilled into Prince William Sound (Ott, 2010; see also "On the Media," June 18, 2010). Walt Parker, who was the head of the Alaska Oil Spill Commission, stated that based on the 1,300 miles of oiled coastline "there is no reason to believe the 11 million gallon estimate" (Bluemink, June 6, 2010).

Similar doubts and suspicions have been cast on BP's estimates of both the size and the location of the spill. Both concerns are especially challenging in this case since the spill occurred at such great depths. Although this spill is unprecedented in size, it is, unlike other major spills, not as obvious on the surface for a couple of reasons. For example, some of the spill may have stabilized in water columns near the sea bed's surface. Also, the extremely large and widespread use of dispersants has made it difficult to measure the size of the spill and its exact location. Dispersants break up the oil into droplets, which become suspended in water and prevent it from coming to the surface.

Most of the early reports were made largely on the basis of BP's own data, a fact which suggests a conflict of interest. The reliance on BP's surveys and data also brings into question the amount of access both the federal government and independent scientists may have to critical information. The government's early reports estimated that the spill was leaking 5,000 barrels a day (Gillis, May 13, 2010). Both environmentalists and independent scientists began challenging this estimate almost from the start. Critics accused the government of using computer models that were not designed for major spills and criticized BP for not employing models that would accurately reflect the true size of the spill.

Florida State University oceanographer Ian R. MacDonald became one of the most outspoken critics. In the *New York Times*, he said that the satellite imagery he was employing suggested that the spill could be four or five times bigger than government estimates (Gillis, May 13, 2010). He also criticized NOAA for failing to institute research that would accurately measure the spill and its effects. He was joined in his criticism by Sylvia Earle, a highly regarded oceanographer and a former chief scientist at NOAA (Gillis, May 13, 2010).

Rick Steiner, the marine scientist involved in the *Exxon-Valdez* oil spill, accused NOAA of a "catastrophic failure" for not accurately analyzing the undersea plumes (Gillis, May 19, 2010), plumes which BP has denied exist. NOAA's estimate was problematic. Figures were based on the amount of oil on the surface. However, as many scientists pointed out, because of the great depth of the leaking oil, the use of dispersants, and the existence of undersea plumes, such calculations were unreliable.

Confusion and uncertainty continued when the Department of Interior's Flow Rate Technical Group released a new report saying that they estimated that two to four times as much oil than the original estimate may be flowing into the Gulf. The group's report said that the new estimate may represent the lower range of "lower bounds." However, the group did not release what the higher bounds were because the higher figure was incalculable because of "known unknowns" and "unknown unknowns" (Froomkin, June 3, 2010).

Representative Edward Markey (D-Mass) was able to shed additional light on the controversy by requesting that BP release its live feed video of the oil spill, a feed that BP had apparently previously denied to both the government and independent scientists. The video feed, however, was of such low quality that information was difficult to gather. Some time later, it was revealed that BP actually had a high resolution live feed video, which they had not made available. When scientists finally viewed the higher-resolution video feed, they concluded the spill was much larger than BP's or the government's estimate (*New York Times*, May 20, 2010). NPR had experts who found that the spill may well be ten times the size of the government estimates. One expert suggested the spill may be as large as 100,000 barrels a day (Harris, May 14, 2010). Not surprisingly, BP disputed the revised estimates.

As of this writing, the debate still rages over the amount of oil spilling into the Gulf of Mexico. The most recent government estimate is that 500,000 to one million gallons of oil per day leaked into the Gulf, which is considerably higher than the original estimates. In keeping with past disasters and the continuing uncertainty about the actual size of the spill, this controversy will no doubt persist for decades and may never ultimately be resolved.

President Obama was reportedly angered by the attempt of independent scientists to gain access to data as well as with criticisms about the reliability of some of the information that his administration has released about the spill (Urbina, May 5, 2010). Such attempts to retard documentation and information gathering unfortunately contribute to the climate of uncertainty that exists in the wake of this catastrophe. Adding to this uncertainty, some critics allege, is President Obama's lack of scientific transparency. The Public Employees for Environmental Responsibility (PEER), on the anniversary of the President's pledge for scientific integrity and transparency to protect science from political interference, reminded the public that his order to the White House Office of Science and Technology Policy states that "federal agency science policy is still being manipulated for political reasons." PEER notes that a year later, no official policy still is in place to preserve scientific integrity. As an example of federal sciences' continued vulnerability to political influence, they note that much of the data that oil companies submit to MMS is not allowed to be circulated for peer review (PEER, July 8, 2010).

Other scientific controversies emerged when Jane Lubchenco, NOAA director, refused to refute BP CEO Tony Hayward's contention that there were no plumes beneath the surface even though NOAA had gathered considerable evidence to the contrary. Jeff Short, a former NOAA official, observed that Hayward's denial of undersea plumes was "right out of the playbook for Exxon and the Exxon saga—which is to say all the right things in public and act deplorably in private." Short contended that Exxon did everything it could to dispute government and independent research with its own questionable research in order to reduce their legal liability (Froomkin, June 3, 2010).

Lubchenco argued that the evidence that had been gathered was "circumstantial." She referred to the oil observed beneath the surface as "anomalies." In a conference call with reporters, she finally relented and said that "it is quite possible that there is oil beneath the surface" (Froomkin, July 13, 2010). According to a report in the *Huffington Post*, some scientists view NOAA's reluctance to make the data public as a move that can only serve BP's interests (Froomkin, July 13, 2010).

Rick Steiner commented on the government's refusal to release scientific findings on the deep sea plumes:

> The fact that they have got this information and they are still not releasing it, that's irresponsible. The public deserves to know this stuff. ... It's infuriating. Here we are in the biggest environmental catastrophe of our time, and the limited amount of science that is being done is being concealed from the public? I mean, c'mon, this is the United States of America. (Froomkin, June 3, 2010)

John T. Hughes, Jr., the director of the worker education training program at the National Institute of Environmental Health Sciences (NIEHS), the institution that in the aftermath of 9/11 raised questions about air quality at the World Trade Center, commented on the ongoing frustration of being unable to gain access to vital data concerning the spill: "The hard part about it is that in a normal response, when the government is doing this, there might be more transparency on the data. In this case when you have BP making the decisions and collecting the data it is harder to have that transparency" (Taylor and Schoof, June 17, 2010).

Frustrated with the federal government's failure to determine the true size and direction of the flowing oil, a team of seventy-seven independent scientists with impeccable credentials and headed by Ira Leifer (a researcher at the Marine Science Institute of the University of California, Santa Barbara), devised a plan to do what NOAA and BP did not accomplish—conduct a rigorous, scientific study to determine the size and consequence of the spill. They have presented BP with an eighty-eight-page plan asking for access and $8.4 million to fund the process. In their mission plan, the authors state: "This scientific mission was developed to address a wide range of critical scientific hypotheses that can only be tested during the actual spill. If we do not seize the moment, then irreplaceable scientific knowledge will be lost to humanity and our response to future accidents greatly diminished" (Inside Higher Education, July 20, 2010). The plan was submitted to BP on June 1, 2010. As of July 1, the scientists had not yet received a response.

BP's attempts to offer researchers generous contracts have also raised concerns about access to scientific information. Reputedly, the oil giant, in anticipation of the hundreds of lawsuits it faces, is seeking to obtain as many expert witnesses as its vast fortune can buy. According to one media account, BP attempted to contract an entire marine sciences department; however, because of the substantial restrictions on the researchers that BP demanded, the scientists declined BP's offer (Raines, July 15, 2010). Not unlike the restrictions Exxon and the TVA placed on its researchers, BP sought to restrict scientists from sharing information with colleagues, talking with the media, and publishing their research. These restrictions are not unlike those that NOAA and other federal agencies have placed on their contractors, with the exception that BP's contracts seem to be, in effect, agreements to join BP's legal team, which seems too restrictive to many. Some researchers being wooed by BP are accepting these conditions, probably to gain access. Carey Nelson, the president of the American Association of University Professors (AAUP), has said that BP's three-year limit on releasing research could delay information that could help protect or restore the environment. Nelson also expressed that much of the research needs to be shared with both the public and the government. His concerns are similar to the concerns of many in the public and academia in the aftermath of the *Exxon-Valdez* spill (*Inside Higher Education* July 20, 2010; *Linkins*, July 16, 2010). According to an article in *Inside Higher Education*, a number of professors have backed out of their agreements with BP in recent weeks (July 20, 2010).

Health Concerns

Deepwater spill cleanup workers face many of the same hazards faced by *Exxon-Valdez* cleanup workers. Unfortunately, in the two decades since the Exxon spill, few long-term studies of the potential long-term effects on workers have been made. The existence of such studies could have provided us with valuable information about how to better protect today's cleanup workers.

A short-term study conducted on the *Prestige* oil spill in Spain (2002) is one of the few of its kind and considered to be the most informative study to date. According to Aubrey Miller, a senior medical advisor to NIEHS, the Institute has contacted the researchers who conducted the investigation for help in designing a research study on the possible health impacts of the Deepwater spill (Raloff, June 17, 2010; Perez-Cadahia, et al., *Env. Health*, 2008; Perez-Cadahia, et al., *Chemosphere*, 2008; Perez-Cadahia, et al., *Mutation*, 2008).

Thus far, hundreds of cleanup workers among the 45,000 workers and volunteers working on the spill have reported symptoms like those of the *Exxon-Valdez* cleanup workers and others: upper respiratory problems, chest pains, skin and eye irritation, nosebleeds, nausea, mental confusion, and stomach problems. A number of environmental health specialists have raised concerns about the potentially adverse effects to both the workers and the public of oil and chemical dispersants. Fortunately, their voices are being heard and their concerns are receiving much more press than did those of us who spoke about similar concerns in the wake of the 1989 spill. The lessons learned, at least by a small but vocal minority in the wake of the *Exxon-Valdez*, Shetland Islands, and *Prestige* spills, and more recently with the health problems associated with 9/11 and Katrina, have ensured that both the media and the public are aware of the inherent dangers that threaten the health of cleanup workers.

As in the *Exxon-Valdez* spill, there is now considerable controversy and concern about the use of dispersants and its effects on the environment and humans. By mid-May, BP had already used 800,000 gallons of these chemicals. Some were injected into the area of the undersea well and sprayed on the surface of the Gulf of Mexico. Public health officials have been concerned about worker exposure to the chemicals, particularly when they are administered by aerial spraying. Some fishermen and workers have told me that they were accidentally sprayed. According to the material data sheets (MSDSs) for the two types of dispersants being used (Corexit 9500 and Corexit 9527A), "no toxicity tests have been conducted on this product." This general statement simply means the products have not been tested for human safety. However, the MSDSs do state that "[h]uman health hazards [are]acute." The Corexit 9527A MSDS warns: "Excessive exposure may cause central nervous system effects, nausea, vomiting, anesthetic or narcotic effects" and that it "may cause liver or kidney effects and/or damage." The government of Great Britain has taken these warnings seriously and banned the use of this dispersant in its waters. The regulatory information section states: "Based on hazard evaluation, none of these substances in this product are hazardous." However, it does make the cautionary disclaimer: "The information in this section is for reference only. It is not exhaustive, and should not be relied upon to take the place of an individualized compliance or hazard assessment. Nalco [the manufacturer] accepts no liability for the use of this information" (Corexit 9527A MSDS). The MSDS also states that inhalation of either product can cause irreversible and permanent damage to red blood cells, liver, and kidneys. Although both MSDSs advise caution to exposure, Hugh Kaufman says, "There is no way to work with this toxic soup without getting exposed" (Thomas, July 28, 2010).

On May 20, 2010, the EPA issued a directive to BP to find a less toxic dispersant within twenty-four hours and begin using it within seventy-two hours (EPA, May 20, 2010). BP responded that the only other federally approved dispersant, Sea Brat 4, that is available in sufficient quantity contains small amounts of a chemical that some federal agencies have identified as an endocrine disruptor and chemicals that could persist in the environment for a long time (Deepwater Horizon, May 22, 2010).

On May 23, the EPA decided that BP should use "as little dispersants as possible" (EPA, May 23, 2010). EPA director Lisa Jackson and then federal on-scene coordinator Rear Admiral Mary Landry directed BP to reduce the use of dispersants by 75 percent from peak usage. According to the EPA website, BP reduced usage 68 percent from that peak. EPA began its own scientific testing of eight dispersants listed on the National Contingency Plan Schedule. Peer-reviewed test results indicate that none of the dispersants tested including Corexit 9500A displayed biologically significant endocrine disrupting activity. The EPA plans to continue by testing the acute toxicity of Louisiana Sweet Crude oil alone and in combination with each of the eight dispersants previously tested (EPA, May 23, 2010).

Eula Bingham, one of the health specialists who visited, raised considerable concern about the health and safety of the Exxon cleanup workers, raised concerns about the health hazards associated with the Deepwater spill: "I think there are community people going out and scooping up tar balls on doing some work that they probably will never get paid by anybody. Who is looking after them? Who is looking at how much exposure they have to these toxic chemicals?" (Hopkins, June 29, 2010). On the web site Pump Handle, Dr. Bingham warned of the potential for long-term health damage and noted that cleanup workers are beginning to report acute symptoms. She has called for their protection both now and in the long term (Hopkins, June 29, 2010).

Despite the demands from health professionals and activists for increased protection for workers, the cleanup crews face considerable threats. As in the aftermath of the Exxon spill, the federal government has provided exemptions from hazardous waste operations and emergency response (HAZWOPER). Under such an exemption, the standard HAZWOPER training requirement of forty hours for workers is not enforced. For some unexplained reason, it took BP over two weeks to set up the minimal four-hour trainings required for workers.

In the early days of the cleanup process, BP had a Vessels of Opportunity agreement form for owners of boats that required the owners to relieve BP of any health and safety training. These forms stipulated that the boat owners agree to work that involved exposure to hazardous chemicals. The United Commercial Fisherman's Association in Louisiana and the Louisiana Environmental Network filed a complaint with the U.S. District Court, Second District, and received a temporary restraining order that required BP to assume responsibility for hazardous chemical exposure safety and to allow boat owners to recover losses. Similar groups throughout the Gulf region have filed suit. Now the training is required to be provided to all Vessels of Opportunity crews. Less than a week later, the Twelfth Circuit District Court ruled in favor of the plaintiffs (Grossman, May 14, 2010).

Wilma Subra, the chemist involved in health issues in the wake of Hurricane Katrina and the Murphy Oil spill and a former consultant to EPA, has accused the EPA of not releasing all the environmental health data they received from BP. She has said that there was considerable anecdotal evidence, based on the 300–400 complaints she had received, that airborne particles of crude oil were being blown ashore under the right wind conditions and threatening public health (Goldenberg, June 20, 2010). Another outspoken and prominent player in previous disasters is Hugh Kaufman, chief investigator for the EPA's ombudsman. Kaufman responded to a statement by Clint Guidry of the Louisiana Shrimp Association that claimed that workers who brought respirators to the site were being threatened with dismissal. Kaufman stated that he had witnessed a similar problem after 9/11: "[T]he administration is down-playing the problem because it saves them money down the line" (Goldenberg, June 20, 2010).

As the controversy about health care concerns increased, questions were raised about BP's exposure monitoring plans and the monitoring data being collected on cleanup workers. A significant part of the data is being gathered by the Center for Toxicology and Environmental Health (CTEH). Although the title suggests an independent research center, this center is actually a for-profit consulting firm. Once again, the government's reliance on data by a contractor hired by the party responsible for the disaster creates a conflict of interest. CTEH was in the center of controversy in the Murphy Oil spill and is the same firm that conducted air monitoring for the TVA after the ash spill, work that generated considerable controversy and was severely criticized by local citizens and the EPA.

This firm also performed toxicity testing on Chinese drywall and found no health risks even though other independent tests had. The drywall has been used predominately in states along the Gulf Coast and was first imported in large quantities after Hurricane Katrina (Grossman, June 28, 2010; Norris, November 29, 2006). The New Orleans branch of Habitat for Humanity built more than 200 homes with the drywall and initially ignored homeowner complaints. According to a *ProPublica* report, the organization has now created a task force to examine the problem (Sapien and Kessler, July 1, 2010). The Consumer Product Safety Commission has found that large amounts of sulfur gasses coming off the board may trigger respiratory problems and corrode wiring, which can cause refrigerators, air conditioners, and other electronic products to fail (*Environmental Health and Engineering*, January 28, 2010; Padgett, March 23, 2009; Allen, October 27, 2009).

As the BP Deepwater spill situation worsened, concern among federal health agencies increased dramatically. One of the key turning points came when David Michaels, the author of *Doubt Is Their Product* and the recently appointed assistant secretary of labor for Occupational Health and Safety, wrote an internal memo complaining about "significant deficiencies in BP's handling of the spill's health-related issues" (Richardson, June 3, 2010; Taylor and Bolstad, May 28, 2010). The memo is a stunning critique of what Michaels perceives as BP's failures to address the health issues of cleanup workers adequately. His memo of May 25, 2010 to the U.S. Coast Guard National Incident Coordinator, now retired Admiral T. W. Allen, includes the following concerns:

> Although we have repeatedly raised these concerns with on-scene BP officials as well as BP corporate leadership, BP has not addressed many of the serious problems in a systematic way. As you are aware, under the National Contingency Plan, OSHA has the responsibility

233

to hold employers, including BP and its contractors, responsible for ensuring the safety of clean-up workers.

OSHA has witnessed numerous deficiencies at several work sites and staging areas through the Gulf Coast region. These are listed below. Our primary concern is that the organizational systems that BP currently has in place, particularly those related to work safety and health training, protective equipment, and site monitoring, are not adequate for the current situation or the projected increase in clean-up operations.

I want to stress that these are not isolated problems. They appear to be indicative of a general systemic failure on BP's part, to ensure the safety and health of those responding to this disaster. Furthermore, BP has not been forthcoming with basic, but critical, safety and health information on injuries and exposures.

This is cause for grave concern and frustration on OSHA's part. Throughout this event, OSHA has been operating in technical support mode and attempting to work cooperatively with BP and its contractors. We are concerned, however, that unless BP takes immediate steps to provide clear directions and oversight to its incident command system for safety, OSHA will need to use its authority to move into enforcement mode in order to ensure the safety of clean-up workers.

Then, in what seemed to be a puzzling turnaround, Michaels stated a few days later that, based on air monitoring tests, cleanup workers were exposed to "minimal" amounts of toxic airborne substances. As a result, OSHA will not require respirators for most workers. This reversal leads us to wonder, On whose science was the decision based?

Without providing any details, Michaels also stated that BP and OSHA were working closely to monitor the worker health safety issues and that, given improvements, there was no need for OSHA enforcement actions. OSHA would focus on assisting BP in complying with health and safety regulations. This turnabout has puzzled the public health community and has left many people wondering what actually transpired between OSHA and BP in the days following Michaels's scathing memo. Why the lack of transparency about the nature of the improvements? What specific agreements were reached between BP and OSHA?

Michaels made another surprising statement in early July. Previously having told *ProPublica* (Chavkin, June 18, 2010) that OSHA was working with BP to improve safety training and that the training would start within a few days, Michaels announced in an interview with *ProPublica* that OSHA was still reviewing the courses (Chavkin, July 2, 2010). An OSHA spokesperson explained this delay in part by the fact that OSHA was satisfied with BP's proposed training improvements. The new course would be eight hours long instead of the previous four hours and would be taught by contractors hired by BP with advice from OSHA. The spokesperson also stated: "OSHA will not approve something that is unsatisfactory, and that is why it is currently working with BP on the third draft of the new curriculum" (Chavkin, July 2, 2010). One wonders what the points of contention are between BP and OSHA.

Increased pressure on federal agencies to protect the public health has lead to some positive measures, including a $10 million project initiated by the Department of Health and Human Services to track potentially adverse acute and long-term oil spill-related health effects. Fourteen thousand spill workers have, to date, volunteered to be a part of the tracking system. Some researchers fear the study is too small and question the reliability of the data, particularly for the BP workers whose illness claims are screened by a private medical service hired by BP. Given the number of lawsuits that are being and will be filed against BP, the situation is ripe for a potential conflict of interest, which causes some critics to be skeptical. Already workers are reportedly reluctant to be tested because of fear of losing their jobs.

Another effort to pursue the long-term health effects of cleanup workers has been launched by the House Committee on Energy and the Environment. Henry Waxman, the Committee chair; Bart Stupak, the Subcommittee's chair on Oversight and Investigation; and Frank Pallone, Jr., Subcommittee chair on Health, have called on Exxon-Mobil to "provide all documents related to the health effects experienced by workers" involved in the *Exxon-Valdez* cleanup efforts (Waxman, et al., July 1, 2010). The Subcommittee on Health is also considering legislation, and the Energy and Commerce Health Subcommittee was slated to consider legislation that would require the federal government to monitor the health of response workers and local residents.

Trailers Redux

The toxic FEMA trailers that became emblematic of inept government, irresponsible policy making and outright deceit in the aftermath of Hurricane Katrina have reappeared. At the end of June, the *New York*

Times broke a story about the trailers' bizarre reappearance in the midst of the Deepwater spill tragedy (Urbina, June 30, 2010). After a protracted scandal, government regulators stipulated the trailers could not be sold as housing. That mandate has not stopped some entrepreneurs: the *Times* story tells of one dealer, Ron Mason, owner of the disaster contracting firm Alpha 1, who sold twenty trailers to companies that contract clean-up workers. In some cases, trailers were sold to the workers themselves. Mason would not reveal to the *Times* whether he informed buyers of the government ban. This is not an isolated incident; other dealers have also sold the banned trailers as housing for spill clean-up workers. The General Services Administration's inspector general has now pursued seven cases in which sellers may not have posted required formaldehyde warnings on the trailers. At the end of the day, some workers return home from a potentially toxic spill to a toxic shelter, exposed to toxins perhaps twenty-four hours a day.

Disturbed by this recent series of events, on July 6, 2010, the House Committee on Energy and Commerce's Subcommittee on Commerce, Trade, and Consumer Protection sent a letter to the administrators of the U.S. Coast Guard and FEMA questioning the ongoing difficulties with the trailers. It included questions about why the agencies did not test the trailers to support their claim that the formaldehyde had off-vented sufficiently that they were no longer hazardous, as well as why, rather than permanently affixing notices to the trailers stipulating that they could not be used for housing, they relied instead on certifications to that effect. The letter also asked what measures the agencies have in place in the event that stipulations of the trailer sales were violated. The agencies were also asked what steps were being taken to address the willful violation of the terms of sale (U. S. House, July 7, 2010).

Government's Role

It has often been said that there is a revolving door between industry and regulatory agencies. At times, apparently, no separation at all exists between the oil industry and federal regulatory agencies. Perhaps the most glaring recent example is that of Gale Norton, former interior secretary (2002–2006), who was heavily criticized for granting three valuable oil shale leases to Royal Dutch Shell shortly before she resigned and accepted a job as a legal advisor to the company giant's oil shale division. The leases were projected to net the corporation hundreds of billions of dollars in profit.

In the early days of the spill, many political commentators and environmentalists blamed both BP and its subcontractors and the previous Bush administration for the events leading up to the Deepwater spill. President Bush came under heavy criticism early in his administration when he appointed Vice-President Cheney to chair a task force to develop a national energy policy. Cheney was criticized for meeting secretively with oil, gas, and energy industry officials whose identity neither he nor President Bush would share. The task force eventually recommended that President Bush issue two executive orders, which Bush promptly issued, only two days after the report was released. One of these orders addressed the task force's finding that there was "too little information regarding the effects governmental regulatory action can have on energy." (Exec. Order 1312, 66 Fed. Reg. 28355 (May 22, 2001). The executive order required agencies to prepare a "Statement of Energy Effects" that would describe potentially harmful effects on energy supply. Curiously, the American Petroleum Institute wrote the draft language for this executive order for the Department of Energy. The Institute's involvement left little doubt of the oil industry's influence on federal policy (U.S. House, Memorandum, July 19, 2010). Secretary Gale Norton was initially responsible for many of oil and gas provisions of Cheney's energy policy.

Critics have also cited the findings of the GAO and other oversight agencies that established that staff members of the Department of the Interior, especially within MMS, had inappropriate relationships with the oil industry, raising questions about the Department's oversight over the oil industry. Under Vice-President Cheney's stewardship, the Bush administration refitted MMS with former industry employees and friends. For example, Steven Giles, a coal industry lobbyist, was appointed deputy secretary, an appointment that provided a tremendous amount of influence in the reshaping of MMS. The remodeling of the agency resulted in weakened safeguards and industry friendly policies. Bobby Maxwell, a former top auditor in MMS, in an interview last year with *Rolling Stone Magazine*, stated, "The oil companies were running MMS in those [Bush] years. Whatever they wanted they got. Nothing was being enforced across the board in MMS" (Dickinson, June 8, 2010). This radical revamping of the agency closely resembled the Bush Administration's fateful reorganization of FEMA and resulted in similar tragic consequences for the Gulf Coast and its residents even after Bush left office.

Investigations revealed that during the Bush administration a lurid culture of accepting gifts and other personal favors from the industry existed within MMS. Perhaps the most damning of these investigations is a report by the Department of Interior's inspector general that investigated allegations about MMS inspectors, including an office in the Lake Charles District of Louisiana, which oversees offshore platform facilities in the Gulf of Mexico (Kendall, May 24, 2010).

According to GAO reports, confidential sources and whistleblowers revealed a litany of illegal, corrupt practices. MMS employees took part in hunting and fishing trips, golf tournaments, Christmas parties, illegal drugs, pornography, and sex with oil industry representatives and employees (Mulkern, September 11, 2008). More serious allegations include the prevalent practice of exempting oil companies from having to provide contingency plans for a major oil spill because they are impractical. Because it was considered by BP to be too expensive, MMS also rejected the idea of requiring a remotely triggered acoustic back-up system that would have closed Deepwater's leaking oil pipe on the seabed floor when the manual shutdown system failed. This back-up system would have also been employed if workers were unable to turn off the switch manually because of a raging fire on the offshore rig, which occurred during the Deepwater Blowout. The switches are said to cost $500,000, which is less than BP's daily cost of operating the Deepwater rig! The MMS's exemption ignored its own internal finding that the remotely triggered switches were "essential" to preventing offshore oil spills (Kennedy, May 5, 2010). That BP voluntarily uses acoustic dead man's switches in the North Sea and in other drilling areas of the world but not in the Gulf of Mexico is highly ironic. Acoustic switches are mandatory for offshore rigs in Brazil and on Norway's North Sea drilling platforms. They are voluntarily used by most major companies along the coasts of Europe.

The credibility of the scandal-ridden MMS was further sullied when GAO and inspector general reports revealed that federal regulators responsible for offshore drilling oversight in the Gulf of Mexico allowed oil industry employees to fill out oil rig inspection reports in pencil. These reports were then erased and filled out by MMS inspectors.

Aside from these objectionable practices, the inspection program appears to exist, for the most part, in name only. Inspector General Mary L. Kendall, in recent Congressional testimony, stated that there are "approximately 60 inspectors for the Gulf of Mexico region to cover nearly 4,000 facilities." She also noted that, ironically, the Pacific Coast has ten inspectors for only twenty-three sites. It is hard not to perceive this asymmetrical balance as somehow an enormous oversight that requires further explanation or as simply intentional (U.S. House, Kendall Testimony, June 17, 2010).

Over the years, other serious allegations about the conflict of interest within MMS have arisen. One of the most notable allegations is an inspector general's report in 2006 outlining a number of serious concerns that called the agency's credibility into question. The report stated that MMS was "fraught with difficulties." Among the deficiencies, MMS was found to have under-collected $700 million dollars in gas royalties, a part of a long legacy within the DOI harking back to the days when the USGS, which was then in charge of collecting royalties, perpetually under-collected. The inspector general noted a climate of distrust and poor communications within the agency. MMS's appearance of conflicts of interest involving the oil and gas industry were also noted. Two years later, another inspector general report concluded that the agency, "modified oil sale contracts without clear criteria, and that modifications appeared to inappropriately benefit the oil companies" (Jackson, May 22, 2010).

However, to suggest that the oil industry enjoyed such cozy relations with the Department of the Interior and MMS only under the Bush administration would be a mistake. While the Bush administration may have significantly reduced government oversight and regulation of the oil industry, the practice of allowing the industry to set standards and have its way in U.S. waters dates back at least several decades. To understand the problem, we have to look back to the early days of offshore drilling in the postwar years. The Outer Continental Shelf Lands Act (OCS) of 1953 assigned responsibility for leasing and regulation of offshore energy and mineral resources outside states waters to the Department of the Interior. Under the Act, the exploration and development of these resources were to be conducted by private industry under competitive bidding and leasing. The Department of the Interior was charged with the responsibility of overseeing and regulating the offshore development. The Department delegated these responsibilities to the U.S. Geological Survey (USGS) and the Bureau of Land Management.

A long practice of the Department of Interior has been to allow the oil companies to set their own rules. The USGS's regulatory decisions were routinely made based on data provided by the oil and gas industry.

This practice became especially controversial during the Santa Barbara oil spill. When a panel recommendation proposed a solution to alleviating the oil gushes off the coast of Santa Barbara, the Department of Interior refused to make public the data or reasoning on which the controversial decision was based. The USGS claimed the data was "proprietary" and could not be released despite the fact that Union Oil, the company responsible for the blowout and spill, had previously signed a clearance for the public release of data (Molotch, 1972). This withholding of information added to the public's uncertainty about what was the best course of action and further eroded the legitimacy of the DOI.

DOI secretary Walter Hickel was pressured by locals and politicians to have the oil companies voluntarily stop drilling until new drilling regulations could be formulated. Hickel, who had close ties to the oil industry, agreed to the demands and announced a voluntary suspension of drilling activities. Unexpectedly, within three days, Hickel produced new regulations, thereby making the offshore drilling moratorium a moot process. To the chagrin of many observers, Hickel's new rules were merely an adoption of the oil companies' already existing rules. His actions made a mockery of the process and fueled public indignation (Sherrill, 540).

George Clyde, a member of the Santa Barbara Board of Supervisors, in a U.S. Senate Subcommittee hearing on Air and Water Pollution of the Committee of Public Works, described Hickel's moratorium maneuvers as a case of "cynicism and pure hypocrisy." Moreover, he alleged that "everyone in the Minerals Division of the Interior department was hell-bent to get drilling underway despite strong opposition from local residents" (Easton, 1972, 66). This is an assessment that many in Southern California shared.

In 1982, Secretary of the Interior James G. Watts, a self-identified zealot who wanted desperately to expand the capture of minerals and energy leasing both on land and offshore, transferred offshore leasing responsibilities from USGS and BLM to the newly created MMS. Needless to say, the move did little, if anything, to overcome the DOI's public image.

While critics may argue whether the Republicans have a cozier relationship with big oil than the Democrats, the DOI's long history of industry favoritism, regardless of which party occupied the White House, should have made it no surprise that the Obama administration's own DOI also suffered from conflict-ridden policies.

Rolling Stone reporter Tim Dickinson's groundbreaking expose on Obama's DOI scandal-ridden relationship with the oil companies laid to rest any disillusions to the contrary. His article made clear that Obama left many of Bush's top appointees in place and either knowingly or unknowingly allowed MMS to rubber-stamp offshore drilling permits without regard to any environmental standards. Perhaps most damningly, Dickinson argued that Obama based his response to the Deepwater spill on "flawed and misleading" BP estimates and had his top players "lowball" the flow rate to a ridiculously low level even after the best science and the historical record made it abundantly obvious that the spill would be worse than the *Exxon-Valdez* spill (Dickinson, June 8, 2010). These egregious actions are from a man who said he would restore scientific integrity to the White House.

Furthermore, in regard to science, just as Bush's administration ignored MMS's own earlier internal warnings about the potential for a major spill and its consequences, the Obama administration, or at least officials within DOI and MMS, ignored a timely warning from one of our nation's top scientific agencies. Only last fall NOAA officials cautioned the DOI that the frequency of offshore oil spills had dramatically increased. The DOI was dangerously underestimating the risks that such a spill would have on coastal residents (*PEER*, October 12, 2009; *PEER*, April 1, 2010; Froomkin, May 3, 2010).

In September 2009, seven months before the fateful spill in the Gulf of Mexico, NOAA raised serious concerns about the Obama administration's Outer Continental Shelf Oil and Gas Strategy and urged the plans be scaled back. Its precautionary recommendations included the creation of an exclusionary zone to prevent leases in the North Aleutians as well as along the Atlantic and Eastern seaboard, that buffer zones should be created in habitats of critical concern, and a recommendation for a moratorium on drilling in the Arctic Ocean until better spill prevention and spill response techniques have been developed. In a memo written by Jane Lubchenco, NOAA administrator, NOAA also recommended that DOI conduct "a more complete analysis of the potential human dimensions of offshore production," a concern that DOI should have been keenly aware of and sensitive to since the *Exxon-Valdez* oil spill and now, even more so, in light of the Deepwater spill.

The Obama administration did heed, eventually, the warnings, somewhat. The sale of leases in the Arctic and Aleutian Bay has been halted, and DOI's revised plan pays more attention to in-depth assessments to environmental impacts on marine habitats. No mention was made, however, of increased concern for the potential impact on coastal populations (Lubchenco, September 21, 2009; Froomkin, May 3, 2010).

To many observers, President Obama's appointment of Ken Salazar as the Secretary of DOI smothered any hopes of true reform within the Department. Ironically, on his appointment of Salazar to the cabinet, Obama said the DOI had for "too long been seen as an appendage to commercial interests" and "[t]hat is going to change under Ken Salazar" (Dickinson, June 8, 2010).

Salazar assumed office and promised to end corruption within the agency. He promised to set new standards within the agency and deal with the few corrupt officials that had tarnished its reputation. While Salazar also demolished the Bush administration plans to open up 300 million acres in Alaska and the Gulf for oil drilling, he nevertheless, according to Dickerson, put up 53 million offshore acres for drilling during the first year in office. Furthermore, he left in office former Bush era appointees who were mired in scandal, including Chris Oynes, the controversial associate director of the offshore energy and minerals management program for the Interior Department's Minerals Management Service. After the Deepwater spill, Oynes was forced to resign. Oynes played a central role in an offshore leasing foul-up that cost taxpayers an estimated $10 billion in lost revenue.

Considering Salazar's previous record in Congress, the appointment seemed incongruous to his past performance and his record of promoting offshore drilling. Dickinson reports that as a senator, Salazar actually criticized President Bush for not making oil companies produce oil faster. He went so far as to steer through Congress the Gulf of Mexico Energy Security Act that opened 8 million acres to drilling.

However, once Salazar became the head of DOI in January 2009, he did undertake some steps toward reform. One of his first acts was to publish a new code of conduct for MMS employees. He also made other modest efforts to reform both DOI and MMS. Along with these reforms, however, he encouraged and endorsed President Obama's proposed expansion of oil and gas drilling off the Outer Continental Shelf, including off the coast of Virginia, and opening new areas in the Gulf of Mexico, which was announced only three weeks before the Deepwater blowout. While the administration has shelved the proposal for the time being, it remains to be seen what the long-term plans are for offshore drilling along the east coast. At the end of June, Salazar did away with the

MMS—but in name only—and announced the creation of the Bureau of Ocean Energy Management Regulation and Enforcement. Whether the renaming and reorganization result in substantial changes remains to be seen. With so many employees from the bureau's previous incarnation still on staff, significant cultural changes seem doubtful.

As of this writing, the oil slick in the Gulf is dissolving, or at least it appears that way. However, there are still many unanswered questions and much uncertainty. The amount of oil beneath the surface and the harm that the dispersants may have caused remain mysteries. As Jane Lubchenco, administrator of NOAA, has stated, "Less oil on the surface does not mean that there isn't oil beneath the surface, however, or that our beaches and marshes are not still at risk" (Gillis and Robertson, July 27, 2010). One might add that many other uncertainties associated with the spill—the health of the cleanup workers, the state of the fisheries, the psychosocial impact on Gulf coastal communities, and a host of other uncertainties—still remain.

In the wake of the *Exxon-Valdez* spill, scientists and social scientists were surprised to discover the unanticipated long-term damage the spill had to Alaska's ecosystem and communities. It is wise for us to remember the disaster continuum in the wake of our most recent catastrophe that still poses a number of uncertainties, some of which may only emerge in the months and years ahead.

Further controversy and uncertainty was generated in early August when the Obama administration claimed that most of the oil from the spill had burned off, dispersed, was skimmed off, or simply evaporated. Press Secretary Robert Gibbs went so far as to make the claim that, not only was the federal response effort responsible for this almost miraculous recovery, but Mother Nature was as well (Froomkin, August 4, 2010). Little did Gibbs know that such a claim bore an eerie resemblance to a similar claim made by a government official in the aftermath of the *Exxon-Valdez* clean-up effort. In April 1990, Secretary of the Interior Manuel Lujan stated, "We can thank God for Mother Nature" (Spence, April 12, 1992). It proved to be an overly optimistic statement that bore little truth in the years to come when research proved that not only had most of the oil not been recovered but that there was unexpected long-term harm to the environment.

The administration's upbeat proclamation was met with skepticism by many researchers. Some scientists, according to a story in the *New*

York Times "attacked the findings and methodology, calling the report premature at best and sloppy at worst" (Gillis and Kaufman, August 4, 2010). One of the most damning comments cited in the *Times* article was by a marine scientist at the University of Georgia, Samantha Joye, who bluntly stated, "If an academic scientist put something like this out, it would be torpedoed into a billion pieces" (Gillis and Kaufman, August 4, 2010). Even if the administration's statement were accurate, as many scientists pointed out, that would leave 26 percent of the oil in the environment, which is hardly a reassuring figure since that would mean that 53 million gallons of oil (an amount several times larger than the *Exxon-Valdez* oil spill) is still present in the Gulf of Mexico!

Even according to scientists involved in writing the report, there was a significant degree of caution and skepticism, since they contended that the figures reported by the administration were based on assumptions and estimates that included a significant margin of error.

Another researcher, Ian McDonald, professor of biological oceanography at Florida State University, went so far as to state that, "There's a lot of ... blue smoke and mirrors in this report. It seems very reassuring, but the data aren't there to actually bear out the assurances that were made" (Fahrenhold, August 5, 2010). In another interview, McDonald stated, "Nobody can reproduce the numbers that the government provides here with the data they are given. So, that's not science. It's something else. Perhaps public relations" (Heller, August 5, 2010). Another highly respected scientist, James Cowan, an oceanographer at Louisiana State University in Baton Rouge, made a more caustic comment on the administration's assertion: "In my mind it is scientifically indefensible" (Schrope, August 10, 2010).

Further controversy ensued when the federal government refused to release to the media, congressional investigators, and independent scientists the research to support the administration's claim. NOAA claimed that the reason why the report could not be released is that it was still being written and it had not yet been peer reviewed! An incredible statement considering that if NOAA's claim is true the administration's statement was premature. NOAA claimed that the early disclosure of the report was the White House's doing rather than NOAA's (Sheppard, August 19, 2010). The statement was revealing, which makes it seem all the more likely that the administration's upbeat claims about the oil was based more on a public relations strategy rather than sound science, a move that further erodes the Obama administration's credibility. Representative Ed Markey (D-MA), whose House oversight committee has been investigating the BP spill, demanded that NOAA turn over to his committee all the data and information of the report. Another member of the committee Representative Darrell Issa (R-CA) made a comment that resonated with many on both sides of the aisle as well as with many researchers: "This is yet another in a long line of examples where the White House's pre-occupation with public relations of the oil spill has superseded the realities on the ground" (Froomkin, August 20, 2010).

Conclusion

The extraordinary series of mishaps that led up to the explosion, fire, and blowout of the Deepwater offshore rig and the events that have ensued underscore all of the recurrent themes addressed in this book. It also demonstrates the painful truth that we have failed to learn from past mistakes. In disaster studies, talking about "lessons learned" from past catastrophes is common. Usually, in tragedies like this one, the discussion turns instead to "lessons not learned." Why do we fail to learn? The disaster throws into sharp relief the fallacy of risk-free offshore oil drilling. But it also raises a more disturbing set of systemic questions about our culture's response to disasters.

First, perhaps the most salient lesson that we failed to learn from the *Exxon-Valdez* oil spill was actually stated emphatically by the Alaska Oil Spill Commission (1990): "Never again should a polluter be placed in charge of the clean-up of a major oil spill." Robust research conducted during the cleanup period demonstrates the myriad ways in which placing Exxon in charge of the cleanup not only undermined the U. S. Coast Guard's ability to oversee the operation, but eventually caused unnecessary, additional harm to both the environment and human population. Why do we continue to put polluters in charge of disaster response?

A second important question is, Why was a corporation that had the worst safety record in U.S. waters and among the worst globally given free rein to conduct offshore drilling? The company, which had incurred the most fines in recent history, was allowed to operate with almost no oversight or regulation at all. Likewise, how was the Department of Interior and the MMS, which was supposed to regulate offshore drilling but which had a long history of allowing the oil industry to regulate itself, allowed to become perhaps the most corrupt agency in the

U.S. government? MMS's level of permissiveness toward the oil industry is astonishing. How is it that BP and other major oil companies were exempted from so many oversights, including the requirement of providing a contingency plan for responding to both regional and major spill events? This question is all the more disturbing since we know from the *Exxon-Valdez* spill that both Exxon and Alyeska failed miserably. How is it that the DOI and MMS did not learn from these earlier shortcomings? Even after it became well-known just how inept and corrupt MMS had become under the Bush administration, why has the Obama administration done so little, until the time of the tragedy, to remedy the situation?

There are many more shocking parallels with the Exxon and other spills that demonstrate apparently how little we have learned in the last two decades. Why has our technological ability to control and clean up oil spills progressed so little, if at all, since 1989? As the Alaska Oil Spill Commission pointed out long ago, "In general, none of the currently available technologies are adequate for these [oil spill] incidents" (1990:100). Uncanny parallels with the *Exxon-Valdez* spill include the uncertainties and controversies surrounding dispersants, *in situ* oil burning, contamination of the food chain, and major concerns about worker health and safety. Certainly, much public attention has been given to this vital public health issue during the current catastrophe. Why has there been so little long-term research into worker health and safety issues? Is this issue being adequately addressed at present? What can be done about the limited ability of workers, the media, and government agencies to gain access to health data?

The lack of transparency of worker health and safety issues during the 1989 spill reveals the conflicts of interest that exist today. Why has BP essentially been allowed to become the gatekeeper of worker health complaints? In the future, we certainly don't want to end up in the position in which we find ourselves today, where Exxon is still refusing to disclose information about worker health and safety two decades after the spill.

These oversights and mismanagements highlight the multitude of issues surrounding the inability of the media, the public, and independent researchers to gain vital information about the spill. The media cannot gain access to oiled beaches; independent researchers cannot gain information about the undersea plumes. Access to information has been a constant theme in all the disasters we have covered. As one scientist stated in Senate hearings about Deepwater, "In the long run, the fear of the unknown only contributes to the stress of the disaster" (Senner, July 27, 2010).

For those of us intimately familiar with the Exxon spill and other disasters, the Deepwater stories considered "breaking news" are old, well-worn narratives. Why did these issues get so little public attention before the Deepwater blowout? How much attention will be given to these issues once the Deepwater spill begins to fade from our nation's attention?

What should we conclude? First, as a social scientist, I believe that we should avoid making the simplistic comparisons that are often made by the media, government officials, corporations, and advocacy groups when they liken one disaster to another. Nonetheless, also as a scholar attuned to cultural and social systems, I believe that depressing similarities do exist. Three decades of disaster research have convinced me that we can learn some general lessons about disasters by analyzing the nature of uncertainty in the wake of catastrophe. As we have witnessed, in times of calamity, corporations, state agencies, social advocacy organizations, and other actors attempt to control disaster narratives by adopting public relations strategies that may either downplay or amplify a sense of uncertainty in order to advance political and policy goals. Obviously, we need to change the way we *conceptualize* disasters and the way we *respond* to them. We cannot afford to repeat these mistakes for decades to come.

Let us now consider some of the things we can do and need to do. Clearly, the paradigmatic shifts we need to make cannot be achieved easily given the sociopolitical landscape of our culture. Also, I do not claim that the changes I recommend will solve all our disaster problems. But they are critical steps.

- We must recognize that disasters are not exceptional events. Disasters, large and small, occur all the time. They are as ubiquitous as the uncertainty that follows in their wake.

- We must stop focusing only on the sensational disasters. We have much to be concerned about and much to learn from everyday disasters as well as major catastrophes. Disasters both large and small reveal a disturbing sameness and have great potential to teach us about our culture.

- We must remember that disasters are grounded in the powerfully political world of social relations. Disasters are not simply socially and politically disruptive; they are *political events* themselves. The control of information, or knowledge, in public discourse, as well as the attempt to control the social production of meaning, is an attempt to define reality. Therefore, it is a distinctly ideological process. In our culture our responses to disasters are primarily shaped by organizations—private, corporate, or governmental—and institutions such as the media, science, and jurisprudence. It is important to remember that none of these is neutral or value-free regardless of their professed mission statements.

• We must abandon our traditionally narrow focus on isolated disaster events and their triggering agent(s). This perspective is too highly condensed. Instead, we must view disasters from the perspective of a long-term continuum. The impact and resonance of the *Exxon-Valdez* spill two decades later has taught us that.

• We must pay close attention to factors that contribute to the effects of a disaster. This analysis shows that we can make, and often do make, the impact of disasters much worse by the way in which we respond to them. All of the cases that we have reviewed illustrate this tendency to make an event worse through our response and our failure to learn from previous mistakes.

• We must change a public discourse that too often defines what is acceptable to say about disasters and usually tries to maintain an emphasis on scientific and technical aspects while avoiding other realms, such as values, ethics, policy, politics, and the opinions of lay people. In the wake of catastrophe, scientific problems are often of vital importance. However, the social problems are often equally important but appear less obvious in the official narratives. Social problems frequently seem more difficult to solve than the technical ones, which may be one of many reasons we overlook or ignore them. Our culture's focus on technological fixes obscures profoundly disturbing social problems and perpetuates the illusion that all we need to do is exercise scientific control over chaos and uncertainty and all will be well. Realistically, science doesn't have the ability to solve these problems. Moreover, this neglect serves the interest of some who may profit from the failure to uncover the underlying socio-political processes that are responsible for the disaster. Certainly, science cannot address the asymmetrical balance of power that lurks in the wings of all disasters. The use of power to amplify uncertainty is a perfect example of how official narratives constrain and overpower alternative narratives.

At the beginning of the twenty-first century, we can look back two decades to the *Exxon-Valdez* oil spill, or four decades back to the Santa

Barbara oil spill, and lament the fact that we ignored major warning signs. Had they been heeded, history might have unfolded differently. We also know that the frequency, severity, and magnitude of disasters is increasing, and that we are making our environment and societies more and more vulnerable to catastrophe. We must ensure that at the end of the twenty-first century, we are not looking back again and realizing that we have again ignored valuable lessons from the past. Now is the time to stop repeating our failures and to start learning from the rich body of information about society, culture, and environment that they have revealed.

References

Ablak, E. 2010. Don't believe every advertisement campaign. *Hurriyet Daily News and Economic Review.* June 23. www.hurriyetdailynews.com/n.php?n=don8217t-believe-every-advertisement-campaign-2010-06-23 (accessed June 23, 2010).

Adams, R. 2010. Gulf oil spill hearing—as it happened: US senators sought answers from the Deepwater Horizon rig's owners and operators in the wake of the Gulf oil spill disaster. On Richard Adams's blog. *Guardian.co.uk.* May 11. http://www.guardian.co.uk/world/richard-adams-blog/2010/may/11/gulf-oil-spill-deepwater-horizon-senate (accessed May 11, 2010).

Agency for Toxic Substances and Disease Registry. 2006. "Formaldehyde sampling of FEMA temporary-housing trailers." Baton Rouge, Louisiana, September-October.

———. 2009. Community meeting about the public health assessment process for the Kingston Coal ash release. Agency for Toxic Substances and Disease Registry (ATSDR) and the Tennessee Department of Health (TDH). Harriman, Tennessee, June 11.

———. 2009. "Vision, Mission, Goals, & Core Values." July 16. http://www.atsdr.cdc.gov/about/mission_vision_goals.html#2 (accessed July 16, 2009).

Aguirre, A. 2010. The reality of life for Louisiana fishermen is an unpalatable choice. *New York Times,* June 9. http://www.nytimes.com/2010/06/10/opinion/10thu4.html?scp=1&sq=The%20reality%20of%20life%20for%20Louisiana%20fishermen%20is%20an%20unpalatable%20choice&st=cse (accessed June 9, 2010).

Alaska Department of Health and Social Services. 1989a. "Public Health Advisory on Crude Oil." April 28.

———. 1989b. "Public Health Advisory on Crude Oil." May 5. Juneau, Alaska.

Alaska Oil Spill Commission Hearings (AOSCH). 1989. Homer, Alaska, July 15.

Alaskan Native Woman. 1990. Interview with author. Tatitlek, Alaska, June 8.

Allen, B. 2003. *Uneasy Alchemy: Citizens and Experts in Louisiana's Chemical Corridor Disputes.* Boston: Massachusetts Institute of Technology.

References

Allen, G. 2009. Toxic Chinese drywall creates a housing disaster. *National Public Radio*, October 27. http://www.npr.org/templates/story/story.php?storyId =114182073 (accessed October 27).

Alpert, B. 2010. Panel chair willing to speed up safety review to reopen rigs: Reilly says interim rules could get drillers working. *Times-Picayune*, June 9. http://www.nola.com/news/t-p/frontpage/index.ssf?/base/news-14/1276064462189870.xml&coll=1 (accessed June 9, 2010).

Alpert, B., and J. Tilove. 2010. Special commission will investigate Gulf of Mexico oil spill. *Times-Picayune*, May 17. http://www.nola.com/news/gulf-oil-spill/index.ssf/2010/05/special_commission_to_inestiga.html (accessed May 17, 2010).

Alsas, S. 1990. Interviewed by William Simeone. Homer, Alaska, March 13.

Alyeska Pipeline Service Company. 1987. "Oil Spill Contingency Plan, Prince William Sound." January. Juneau, AK: Alyeska.

American Civil Liberties Union (ACLU). 2010. ACLU reminds law enforcement to respect media and public access to BP oil spill. *Louisiana American Civil Liberties Union*, June 28. http://www.laaclu.org/newsArchive.php?id=378#n378 (accessed June 28, 2010).

American Petroleum Industry. 2009. Offshore access to oil and natural gas resources. Washington, DC. www.api.org (accessed July 12, 2010).

Anderson, A. 1990. Interview with author. Homer, Alaska, May 16.

Andrews C. 1993. Interview with author. Lerwick, Shetland Islands, February 12.

Appalachian Voices. 2009. Preliminary tests find high levels of toxic chemicals in Harriman TN fly ash deposits. *Appalachian Voices*. Press release, January 1.

Appleton, B. 1994. Piper Alpha. In *Lessons from disaster: How organisations have no memory and accidents recur.* Edited by T. Kletz, 174–84. London: Institute of Chemical Engineers.

Associated Press. 2008. Scientists closely examine FEMA trailers. *New York Times*, February 16. http://www.nytimes.com/aponline/us/AP-Toxic-Trailers.html (accessed February 17, 2008).

Bachrach, P., and M. Baratz. 1970. *Power and poverty: Theory and practice.* New York: Oxford University Press.

Baker, P. 2010. Obama gives a bipartisan commission six months to revise drilling rules. *New York Times*, May 22. http://query.nytimes.com/gst/fullpage.html?res=9D05E1D61F30F930A15756C0A9669D8B63&scp=4&sq=%22Peter+Baker%22&st=nyt (accessed May 22, 2010).

Bankoff, G., G. Frerks, and D. Hilhorst., eds. 2004. *Mapping vulnerability: Disasters, development, and people.* Sterling, VA: Earthscan.

Baram, M. 2010. Interior secretaries under Bush and Obama exposed by house panel for lax oversight of oil drilling. *Huffington Post*, July 20. http://www.huffingtonpost.com/2010/07/19/interior-secretaries-unde_n_652110.html (accessed July 21, 2010).

Barker, S. 2009a. As professionals weigh in, debate over fly ash spill continues. *Knox News Sentinel*, July 12. http://www.knoxnews.com/news/2009/jul/12/debate-over-cause-of-ash-spill-continues/ (accessed July 14, 2009).

——. 2009b. TVA approves test of landfills for ash: Sites will receive 5–10 truckloads of sludge from spill. *Knox News Sentinel*, July 2. http://www.knoxnews.com/news/2009/jul/02/tva-approves-test-of-landfills--for-ash/ (accessed July 2, 2009).

Baugher, T. 1990. Interview with author. Homer, Alaska, August 9.

Beaumont, P. 2010. Louisiana oil spill: Toxic chemical fear over BP's clean-up efforts. *Guardian.co.uk*. May 16. http://www.guardian.co.uk/environment/2010/may/16/louisiana-oil-spill-toxic-chemical-bp (accessed May 16, 2010).

Beck, U. 1992. *Risk society: Towards a new modernity*. London: Sage Publications.

——. 1999. *World risk society*. Cambridge: Polity Press.

Begley, S. 2010. What the spill will kill. *Newsweek*, June 4. http://services.newsweek.com/id/238620 (accessed June 4, 2010).

Bennett, J. 2006. It's "Incompetence": An environmental expert fears that Gulf Coast residents and volunteers exposed to deadly toxins could suffer health effects similar to those of 9/11 workers. *MSNBC.com*, August 24. http://www.msnbc.msn.com/id/14497763/site/newsweek/page/4/ (accessed August 24, 2006).

Benson, P. 1990. Interview with author. Homer, Alaska, July 31.

Bird Morgue Worker. 1990. Interview with author. Homer, Alaska, August 9.

Bisek, C. 2006. Interview with author, New Orleans, Louisiana, April.

Blucher, J. 1990. Spill exhibits lose venue: Government questioned "property." *Anchorage Daily News*, October 23, B1, B3.

Bluemink, E. 2010. Activists still challenge official estimate of Exxon Valdez spill. *McClatchy Newspapers*, June 6. http://www.mcclatchydc.com/2010/06/06/95398/activists-still-challenge-estimate.html (accessed June 6, 2010).

Bodett, T. 1990. Interview with author. Homer, Alaska, August 8.

Bolstad, E. 2010. Oil spill taking toll on BP's credibility—and the government's. *McClatchy Washington Bureau*, May 29. http://www.mcclatchydc.com/2010/05/29/95061/oil-spill-is-taking-a-toll-on.html (accessed May 29, 2010).

Borkowski, L. 2010. Sequestered silence: Oil cleanup workers' health. *Pump Handle*, May 10. http://thepumphandle.wordpress.com/2010/05/10/sequestered-science-oil-cleanup-workers-health/ (accessed May 10, 2010).

Bowen, M. 2008. *Censoring science: Inside the political attack on Dr. James Hansen and the truth of global warming*. New York: Plume.

Broder, J. M. 2010. U.S. to split up agency policing the oil industry. *New York Times*, May 11. http://query.nytimes.com/gst/fullpage.html?res=9B01EFD7173BF931A25756C0A9669D8B63&sec=&spon=&pagewanted=all (accessed May 11, 2010).

References

Broder, J. M., and M. Luo. 2010. Reforms slow to arrive at drilling agency. *New York Times*, May 30. http://www.nytimes.com/2010/05/31/us/politics/31drill. html?scp=1&sq=%22Reforms%20slow%20to%20arrive%20at%20drilling%20 agency%22&st=cse (accessed May 30, 2010).

Brown, M. 2010a. Gulf oil spill: Media access "slowly being strangled off." *Huffington Post*, May 29. www.huffingtonpost.com (accessed May 30, 2010).

———. 2010b. Underwater oil plumes disputed by BP CEO Tony Hayward. *Huffington Post*, May 30. http://www.huffingtonpost.com/2010/05/30/underwater-oil-plumes-dis_n_595015.html (accessed May 31, 2010).

Brown, P., and E. J. Mikkelsen. 1990. *No safe place: Toxic waste, leukemia, and community action.* Berkeley: University of California Press.

Browne, M. 1990. $51,260 to rehabilitate an otter was worth it, researchers say. *New York Times*, October 2, C4.

Brunker, M. 2008. CDC tests confirm FEMA units are toxic: Agency to relocate Gulf Coast residents because of formaldehyde fumes. *MSNBC.com*, February 14. http://www.msnbc.com/id/23168160 (accessed February 17, 2008).

Burger, J., ed. 1994. *Before and after an oil spill: The Arthur kill.* New Brunswick, NJ: Rutgers University Press.

Bushnell, S., and S. Jones. 2009. *The spill: Personal stories from the Exxon Valdez disaster.* Kenmore, WA: Epicenter Press.

Button, G. 1993. Social conflict and the formation of emergent groups in a technological disaster: The Exxon Valdez oil spill and the response of the residents in the area of Homer, Alaska. Ph.D. dissertation, Brandeis University.

———. 1994a. The apocalypse postponed: The media response to the Shetland Island oil Spill. Paper on file with the author.

———. 1994b. The *Exxon-Valdez* oil spill: An environmental or human disaster? Paper presented at the Annual Meeting of the American Public Health Association, November 1, 1994, in Washington, DC.

———. 1999. The negation of disaster: The media response to oil spills in Great Britain. In *The angry earth: Disaster in anthropological perspective*, edited by A. Oliver-Smith and S. M. Hoffman. New York: Routledge.

———. 2002. Popular media reframing of man-made disasters: A cautionary tale. In *Catastrophe and culture: The anthropology of disaster*, edited by S. M. Hoffman and A. Oliver-Smith. Santa Fe: School of American Research Press.

Button, G., and A. Oliver-Smith. 2008. Disaster, displacement, and employment: Distortion of labor markets during post-Katrina reconstruction. In *Capitalizing on catastrophe: Neoliberal strategies in disaster reconstruction*, edited by N. Gunewardena and M. Schuller. New York: AltaMira Press.

Button, G., and K. Peterson. 2009. Participatory action research: Community partnership with social and physical scientists. In *Anthropology and climate change: From encounters to actions*, edited by S. A. Crate and M. Nuttall. Walnut Creek, CA: Left Coast Press.

References

Campbell, D., D. Cox, J. Crum, K. Foster, P. Christie, and D. Brewster. 1993. Initial effects of the grounding of the tanker Braer on health in the Shetland. *BMJ* 307: 1251–55.

Campbell, D., D. Cox, J. Crum, and A. Riley. 1994. Later effects of grounding of tanker Braer on health in Shetland. *BMJ* 309: 773–74.

Cardinale, M. 2010. Fears grow over oil spill's long-term effects on food chain. *Guardian.co.uk*, June 1. http://www.guardian.co.uk/environment/2010/jun/01/bp-oil-spill-wildlife (accessed June 1, 2010).

Carson, W. 1982. *The other price of Britain's oil: Safety and control in the North Sea.* New Brunswick, NJ: Rutgers University Press.

CBS News. 2010. Shrimper douses herself with oil in spill protest. *CBSNews.com*, June 9. http://www.cbsnews.com/8301-503544_162-20007240-503544.html (accessed June 9, 2010).

Center for Biological Diversity. 2010. Gulf disaster: End offshore oil drilling now. *Center for Biological Diversity*, June 29. http://www.biologicaldiversity.org/programs/public_lands/energy/dirty_energy_development/oil_and_gas/gulf_oil_spill/index.html (accessed June 29, 2010).

Centers for Disease Control. n.d. "Indoor air quality and health in fema temporary housing: For healthcare providers." Pamphlet. Centers for Disease Control.

———. n.d. "Indoor air quality and health in FEMA temporary housing: For trailer residents." Pamphlet. Centers for Disease Control.

———. 2009. "Evaluation of mitigation strategies for reducing formaldehyde concentrations in unoccupied Federal Emergency Management Agency-owned travel trailers." November 10.

Chavkin, S. 2010a. OSHA director: Offshore cleanup workers will get more training. *ProPublica,* June 18. http://www.propublica.org/article/osha-director-offshore-cleanup-workers-will-get-more-training (accessed June 18, 2010).

———. 2010b. Two weeks later, new safety training for Gulf workers has yet to begin. *ProPublica*, July 2. http://www.propublica.org/article/two-weeks-later-new-safety-trainings-for-gulf-workers-have-yet-to-begin (accessed July 2, 2010).

Chelsea Green. 2009. Riki Ott: These big corporations prey on people's trust. March 24. http://www.chelseagreen.com/content/riki-ott-these-big-corporations-prey-on-peoples-trust/ (accessed March 24, 2009).

Cherniack, M. 1986. *The hawk's nest incident: America's worst industrial disasters.* New Haven, CT: Yale University Press.

Chittum, R. 2010. BP journalists and their greatest propaganda hits. *Columbia Journalism Review*, June 22. http://www.cjr.org/the_audit/bp_journalists_and_their_great.php (accessed June 22, 2010).

Choate, B. 1990. Interview with author. Homer, Alaska, September 12.

Christianson, A. 1990. Interview with author. Homer, Alaska, August 11.

Christmas, R. 1990. Interview with author. Homer, Alaska, April 18.

Citizens Commission Hearings on the *Exxon-Valdez* Oil Spill. 1989. "The Day the Water Died." Anchorage, Cordova, Homer, Kodiak, and Old Harbor, Alaska, November.

Clarke, L. 1989. *Acceptable risk? Making decisions in a toxic environment.* Berkeley: University of California Press.

———. 1999. *Mission impossible: Using fantasy documents to tame disaster.* Chicago: University of Chicago.

———. 2006. *Worst cases: Terror and catastrophe in the popular imagination.* Chicago: The University of Chicago Press.

Clifford, J., and G. Marcus, eds. 1986. *Writing culture: The poetics and politics of ethnography.* Berkeley: University of California Press.

Coates, P. 1991. *The trans-Alaskan pipeline.* LeHigh, PA: LeHigh University Press.

Cohen, A. P. 1987. *Whalsay: Symbol, segment and boundary in a Shetland Island community.* Manchester, UK: University of Manchester Press.

Collins, J. 1992. The potential for right-to-know legislation in Canada. *International Journal of Mass Emergencies and Disaster* 10(2): 349–64.

Comaroff, J., and S. Roberts. 1981. *Rules and processes.* Chicago: University of Chicago Press.

Cooper, C., and R. Block. 2006. *Disaster: Hurricane Katrina and the failure of Homeland Security.* New York: Times Books.

Coppens, P. 2008. Wikipedia lies: Online disinformation & propaganda. January 20. http://educate-yourself.org/cn/wikipedialies20jan08.shtml. (accessed January 20, 2008).

Corexit 9500; MSDS. 2005. Nalco: Sugar Land, Texas. June 14. http://www.nalco.com/documents/9500A_MSDS.pdf (accessed on May 5, 2010).

Corexit (R) EC9527A; MSDS. 2008. Nalco: Sugar Land, Texas. October 15. http://www.deepwaterhorizonresponse.com/posted/2931/Corexit_EC9527A_MSDS.539295.pdf (accessed on May 5, 2010).

Coughlin, W. 1992. Valdez cleanup linked to ailments. *Boston Globe*, May 10.

Cowan, E. 1968. *Oil and water: The Torrey Canyon disaster.* Philadelphia and New York: J. B. Lippincott Company.

Cox, D. 1993a. The *Braer* incident. Health impact[s]: Immediate impact-response. Paper presented at the Shetland Islands Marine Environment Conference, March 31, Lerwick.

———. 1993b. Interview with author. Lerwick, Shetland Islands. February 8.

Cranor, C. F. 2006. *Toxic torts: Science, law, and the possibility of justice.* New York: Cambridge University Press.

Crouch, S. R., and S. Kroll-Smith. 2000. Environmental movements and expert knowledge: Evidence for a new populism. In *Illness and the environment: A reader in contested medicine*, edited by S. Kroll-Smith, P. Brown, and V. J. Gunter. New York: New York University Press.

References

Cuddehe, M. 2010. A spy in the Jungle. *Atlantic.com,* August 2. http://www.the-atlantic.com/international/archive/2010/08/a-spy-in-the-jungle/60770/ (accessed August 2, 2010).

Das, 2000. Suffering, legitimacy, and healing: The Bhopal case. In *Illness and the Environment: A Reader in Contested Medicine,* edited by S. Kroll-Smith, P. Brown, and V. J. Gunter. New York: New York University Press.

Daubert v Merrell Dow Pharmaceuticals. 1993. 509 U.S. 579.

Davidson, A. C. 1990. *In the wake of the EXXON Valdez: The devastating impact of the EXXON Valdez oil spill.* San Francisco: Sierra Club Books.

Davis, D. 2002. *When smoke ran like water: Tales of environmental deception and the battle against pollution.* New York: Basic Books.

———. 2007. *The secret history of the war on cancer.* Philadelphia: Basic Books.

Davis, M. 2009. Davis: Are birds threatened by Kingston fly ash spill? *Knox News Sentinel,* June 28. http://www.knoxnews.com/news/2009/jun/28/are-birds-threatened-by-kingston-fly-ash-spill/ (accessed June 29, 2009).

Deepwater Horizon Unified Command official Web site. 2010. Deepwater Horizon response. http://www.deepwaterhorizonresponse.com/go/doc/2931/559595/ (accessed May 23, 2010).

DeLillo, D. 1984. *White noise.* New York: Viking.

DeMain, J. 2010. The environmentalists and deepwater drilling. *A Hard Day's Blog,* June 25. http://open.salon.com/blog/jeanette_d/2010/06/16/the_environmen-talists_and_deepwater_drilling (accessed June 25, 2010).

Democracy Now! 2010a. Journalist exposes how private investigation firm hired by Chevron tried to recruit her as a spy to undermine $27B Suit in Ecuadorian Amazon. *DemocracyNow.org,* August 16. http://www.democracynow.org/2010/8/16/journalist_exposes_how_private_investigation_firm (accessed August 16, 2010).

———. 2010b. BP oil spill confirmed as worst in US history; Environmental groups challenge continued oil operations in Gulf excluded from New Moratorium. *DemocracyNow.org,* May 28. http://www.democracynow.org/2010/5/28/bp_oil_spill_confirmed_as_worst (accessed May 28, 2010).

———. 2010c. *Rolling Stone's* Tim Dickinson on the inside story of how Obama let the world's most dangerous oil company get away with murder. *DemocracyNow.org,* June 11. http://www.democracynow.org/2010/6/11/rolling_stones_tim_dickinson_on_the (accessed June 11, 2010).

———. 2010d. Scientist: BP well could be leaking 100,000 barrels of oil a day. *DemocracyNow.org,* June 9. http://www.democracynow.org/2010/6/9/scientist_bp_well_could_be_leaking (accessed June 9, 2010).

Desimone, B. 2010. In reporting on oil spill, limits persist on media access in the Gulf. *PBS,* June 30. http://www.pbs.org/newshour/rundown/2010/06/access-hard-to-come-by-in-reporting-on-health-in-the-gulf.html (accessed on June 30, 2010).

References

Dewan, S. 2008a. At plant in coal ash spill toxic, deposits by the ton. *New York Times,* December 30. http://www.nytimes.com/2008/12/30/us/30sludge.html? (accessed September 13, 2009).

———. 2008b. Tennessee ash flood larger than initial estimate. *New York Times,* December 27. http://www.nytimes.com/2008/12/27/us/27sludge.html?_r=1&scp =1&sq=Tennessee%20Ash%20Flood%20Larger%20Than%20Initial%20 Estimate&st=cse (accessed January 4, 2009).

———. 2009a. Holdouts test aid's limitations as FEMA shuts. *New York Times,* June 7.

———. 2009b. Katrina victims will not have to vacate trailers. *New York Times,* June 4. http://www.nytimes.com/2009/06/04/us/04trailers.html (accessed June 8, 2009).

———. 2009c. Ready or not, Katrina victims lose temporary housing. *New York Times,* May 8. http://www.nytimes.com/2009/05/08/us/08trailer.html (accessed June 21, 2009).

Dickinson, T. 2010a. BP's next disaster. *Rolling Stone,* June 22. http://www.rolling-stone.com/politics/news/17390/120130 (accessed June 22, 2010).

———. 2010b. The spill, the scandal and the president. *Rolling Stone,* June 8. http:// www.rollingstone.com/politics/news/17390/111965 (accessed June 8, 2010).

DiPerna, P. 1985. *Cluster mystery: Epidemic and the children of Woburn, MASS.* St. Louis: Mosby Company.

Dlouhy, J. 2010. Oil CEO's criticize BP well, but have identical response plans. *Houston Chronicle,* June 15. http://www.chron.com/disp/story.mpl/business/ 7054419.html (accessed June 15, 2010).

Dold, C. 1991. Otter rescue in 1989 is questioned anew on ecological grounds. *New York Times,* December 31, C4.

Doms, N. 2010. The toxicity of Corexit in BP oil cleanup: A hazard for workers. *HULIQ.com.* June 27. http://www.huliq.com/9990/toxicity-corexit-bp-oil-cleanup-hazard-workers (accessed June 27, 2010).

Douglas, M., and A. Wildavsky. *Risk and culture: An essay on the selection of tech-nological and environmental dangers.* Berkeley: University of California Press.

Duckworth, P. 1986. Psychological problems arising from disaster work. *Stress Medicine* 2: 315–23.

Dumanoski, D. 1989. Volunteers are cleaning up what Exxon left behind. *Boston Globe,* September 17, 1.

Dye, L. 1971. *Blowout at platform A: the crisis that awakened a nation.* Garden City, NY: Doubleday and Company.

Easton, R. 1972. *Black tide: The Santa Barbara oil spill and its consequences.* New York: Delacorte Press.

Eaton, L. 2008. FEMA trailers found toxic: Agency to hasten relocation of hurri-cane victims. *Boston Globe,* February 15. http://www.boston.com/news/nation/ articles/2008/02/15/fema_trailers_found_toxic?mode=PF (accessed February 17, 2008).

References

Edelman, M. 1977. *Political language.* New York: Academic Press.

Edelstein, M. 1988. *Contaminated communities.* Boulder, CO: Westview Press.

Egginton, J. 1980. *The poisoning of Michigan.* New York: Norton.

Eisenberg, N. 2010. Onshore drilling disasters waiting to happen: An interview with "Gasland" director Josh Fox. *Nation,* June 17. http://www.thenation.com/article/36385/onshore-drilling-disasters-waiting-happen-interview-gasland-director-josh-fox (accessed June 17, 2010).

Engelberg, S. 2010. *ProPublica* photographer detained by BP and local police. *ProPublica,* July 3. http://www.propublica.org/article/photographer-detained-briefly-by-bp-and-local-police (accessed on July 3, 2010).

Enhorn, S. 2009. Interview with author. Roane County, Tennessee, May.

Environmental Health and Engineering, Inc. 2010. Final report on an indoor environmental quality assessment of residences containing Chinese drywall. Prepared for Lori Saltzman, M.S. Director, Division of Health Sciences U.S. Consumer Product Safety Commission, January 28. http://www.cpsc.gov/library/foia/foia10/os/51homeFinal.pdf (accessed January 28, 2010).

Environmental Research Foundation. 1992. What has gone wrong? Part 1: Congress creates a monster: The ATSDR. *Rachel's Hazardous Waste News,* no. 292 (July 1). http://www.rachel.org/en/newsletters/rachels_news/ (accessed January 15, 2010).

Estes, J. 1991. Catastrophes and conversation: Lessons from the sea otters and the *Exxon Valdez. Science* 254: 1596.

Evans, B. 2008a. FEMA limits formaldehyde in trailers. *Associated Press,* April 12. http://ap.google.com/article/ (accessed April 14, 2008).

———. 2008b. Scientist: CDC bosses ignored warning. *Associated Press,* April 1. http://ap.google.com/article/ (accessed April 2, 2008).

Evans, L. 2009. E-mail message to author, May 18.

Ewen, S. 1996. *PR! A social history of spin.* New York: Basic Books.

Fagin, D., and M. Lavelle. 1999. *Toxic deception: How the industry manipulates science, bends the law and endangers your health.* Monroe, ME: Common Courage Press.

Fahrenthold, D. 2010. Scientists question government team's report of shrinking gulf oil spill. *Washington Post,* August 5. http://www.washingtonpost.com/wp-dyn/content/article/2010/08/04/AR2010080407082.html (accessed September 16, 2010).

Fairhall, D., and P. Jordan. 1980. *The wreck of the "Amoco Cadiz": The story of the greatest ecological disaster at sea and what it bodes for the future.* New York: Stein and Day.

Fall, J. 1990. Subsistence after the Spill: Uses of fish and wildlife in Alaska native villages and the *Exxon Valdez* oil spill. Paper presented at the Annual Meeting of the American Anthropological Association, New Orleans, Louisiana. November.

References

Federal Emergency Management Agency. 2008. USA: Myths and facts about FEMA housing following Katrina. Press release, May 26. http://www.fema.gov/news/newsrelease.fema?id=43544 (accessed May 28, 2008).

Feit, H. A. 1991. Metaphors of nature and the love of animals: Comparisons of animal rights supporters and James Bay Cree hunters. Paper presented at the Annual Meeting of the American Anthropological Association, Chicago, Illinois. November.

Fernandez, D. 2008. Toxic trailers redux: When did FEMA know? Newly found documents show OSHA detected dangerous levels of formaldehyde in trailers used to house Katrina evacuees as early as 2005—but FEMA mass distributed them anyway. *Mother Jones*, March 25. http://www.motherjones.com/news/update/2008/03/fema-toxic-trailers-2.html. (accessed April 1, 2008).

Finch, S. 2007. Hearing set for Murphy Oil spill. January 1. http://www.nola.com/news/t-p/metro/index.ssf?/base/news-19/1167633095213450.xml&coll=1.

Fingas, M. 2001. *The basics of oil spill cleanup*. Boca Raton, FL: Lewis Publishers.

Flessner, D., and P. Sohn. 2008. Tennessee Valley Authority boosts estimate from coal as spill. *Chattanooga Times Free Press*, December 27. http://www.timesfreepress.com/news/2008/dec/27/tennessee-valley-authority-boosts-estimate-coal-as/ (accessed December 27, 2008).

Fogg, R. 1993. Health study results "reassuring." *Shetland Times*, February 26.

Ford, K. 2006. New study exposes Exxon Chalmette refining's accident rate. *Global Community Monitor*. St. Bernard Citizens for Environmental Quality / Louisiana Bucket Brigade Press release, November 1. http://www.shellfacts.com/article.php?id=480 (accessed March 3, 2008).

Fortun, K. 2001. *Advocacy after Bhopal: Environmentalism, disaster, new global orders*. Chicago: The University of Chicago Press.

Fountain, H. 2010. Advances in oil spill cleanup lag since *Valdez*. *New York Times*, June 24. http://www.nytimes.com/2010/06/25/us/25clean.html?scp=1&sq=Advances%20in%20oil%20spill%20cleanup%20lag%20since%20Valdez&st=cse (accessed June 24, 2010).

Fountain, H., and T. Zeller, Jr. 2010. Panel suggests signs of trouble before rig explosion. *New York Times*, May 25. http://www.nytimes.com/2010/05/26/us/26rig.html?scp=1&sq=%22Panel%20suggests%20signs%20of%20trouble%20before%20rig%20explosion%22&st=cse (accessed May 25, 2010).

Freudenberg, W. 1988. Perceived risk: Real risk: Social science and the art of probabilistic risk assessment. *Science* 244: 127.

———. 2005. Seeding science, courting conclusions: Reexamining the intersection of science, corporate cash, and the law. *Sociological Forum*. Vol. 20, No. 1, March.

Freudenburg, W., and R. Gramling. 1994. *Oil in troubled waters: Perceptions, politics, and the battle over offshore drilling*. Albany: State University of New York Press.

———, R. Gramling, S. Laska., and K. Erikson. 2009. *Catastrophe in the making: The engineering of Katrina and the disasters of tomorrow*. Washington, DC: Island Press.

References

Froomkin, D. 2010a. BP's response plan was a joke, group charges. *Huffington Post*, May 24. http://www.huffingtonpost.com/2010/05/24/bps-response-plan-was-a-j_n_587846.html (accessed May 24, 2010).

———. 2010b. Gulf oil spill: BP's continued denial of underwater plumes provokes ridicule. *Huffington Post*, June 10. http://www.huffingtonpost.com/2010/06/09/gulf-oil-spill-bps-contin_n_606819.html (accessed June 10, 2010).

———. 2010c. Gulf oil spill: BP's poor record so far dulls hope for future. *Huffington Post*, June 11. http://www.huffingtonpost.com/2010/06/11/gulf-oil-spill-bps-poor-r_n_608114.html (accessed June 11, 2010).

———. 2010d. Gulf oil spill: Federal estimates again eclipsed by reality. *Huffington Post*, June 8. http://www.huffingtonpost.com/2010/06/08/gulf-oil-spill-federal-estimate_n_605095.html (accessed June 9, 2010).

———. 2010e. Gulf oil spill: Latest federal government estimate still understates oil flow. *Huffington Post*, June 3. http://www.huffingtonpost.com/2010/06/03/gulf-oil-spill-latest-fed_n_599615.html (accessed June 4, 2010).

———. 2010f. NOAA hoarding key data on oil spill damage. *Huffington Post*, July 13. http://www.huffingtonpost.com/2010/07/13/noaa-hoarding-key-data-on_n_645031.html (accessed July 13, 2010).

———. 2010g. NOAA warned Interior it was underestimating threat of serious spill. *Huffington Post*, May 3. http://www.huffingtonpost.com/2010/05/03/noaa-warned-interior-was_n_561615.html (accessed May 6, 2010).

———. 2010h. Obama's options: What he can and should say about the oil spill today. *Huffington Post*, May 27. http://www.huffingtonpost.com/2010/05/27/obamas-options-what-he-ca_n_590856.html (accessed May 27, 2010).

———. 2010i. Administration overly optimistic about fate of spilled oil. Video. *Huffington Post*. August 4. http://www.huffingtonpost.com/2010/08/04/administration-overly-opt_n_671090.html (accessed August 4, 2010).

———. 2010j. Questions mount about White House's overly rosy report on oil spill. *Huffington Post*. August 20. http://www.huffingtonpost.com/2010/08/20/overly-rosy-report-on-oil_n_688142.html (accessed August 20, 2010).

Frost, G. 1990. Putting a spin on the spill: Numbers and images become chess pieces in battle to control coverage. *Anchorage Daily News*, March 27, A1, A4.

Fuller, J. G. 1977. *The poison that fell from the sky*. New York: Random House.

Gambrell, J. 2008. FEMA to use trailers after tornadoes. *Associated Press*, February 13. http://ap.google.com/article/ (accessed February 18, 2008).

Gerber, G. 2009. FEMA trailers: The gift that keeps on taking year after year. *RV E-News*, July 13. http://www.toxictrailers.com. (accessed July 26, 2009).

Gibbs, L. 1982. *Love Canal: My story*. Albany: State University of New York Press.

Giddens, A. 1991. *Modernity and self-identity*. Stanford, CA: Stanford University Press.

———. 2000. *Runaway world: How globalization is reshaping our lives*. New York: Routledge.

References

Gill, C., F. Booker, and T. Soper. 1967. *The wreck of the "Torrey Canyon."* Newton Abbot, Devon, UK: David and Charles.

Gillis, J. 2010a. Scientists fault lack of studies over Gulf oil spill. *New York Times,* May 19. http://www.nytimes.com/2010/05/20/science/earth/20noaa.html?scp=1&sq=%22Scientists%20fault%20lack%20of%20studies%20over%20Gulf%20oil%20spill%22&st=cse (accessed May 19, 2010)

———. 2010b. Size of oil spill underestimated, scientists say. *New York Times,* May 13. http://www.nytimes.com/2010/05/14/us/14oil.html?ref=justin_gillis (accessed May 13, 2010).

Gillis, J., and L Kaufman. 2010. Oil spill calculations stir debate on damage. August 4. *New York Times.* http://www.nytimes.com/2010/08/05/us/05oil.html (accessed August 4, 2010).

Gillis, J., and C. Robertson. 2010. On the surface, Gulf oil spill is vanishing fast; Concerns stay. *New York Times,* July 27. http://www.nytimes.com/2010/07/28/us/28spill.html (July 27, 2010).

Gitlin, T. 1980. *The whole world is watching.* Berkeley: University of California Press.

Goddard, J. 2010. BP accused of using Gulf of Mexico as a "toxic testing-ground." *CommonDreams.org,* May 15. http://www.commondreams.org/headline/2010/05/15-6 (accessed May 15, 2010).

Goldenberg, S. 2010a. BP accused of killing endangered sea turtles in cleanup operation: Environmentalists press Obama administration to put a halt to BP's "burn fields" to dispose of oil from Gulf spill. *CommonDreams.org,* June 25. http://web001.commondreams.org/headline/2010/06/25-9 (accessed June 25, 2010).

———. 2010b. BP will pay "many billions of dollars in fines" for oil spill, White House warns. *Guardian.co.uk,* June 7. http://www.guardian.co.uk/environment/2010/jun/07/bp-oil-spill-fines-government (accessed June 7, 2010).

———. 2010c. Gulf oil spill is public health risk, environmental scientists warn. *Guardian.co.uk,* May 28. http://www.guardian.co.uk/environment/2010/may/28/bp-gulf-oil-spill-pollution (accessed May 28, 2010).

———. 2010d. Tony Hayward's worst nightmare? Meet Wilma Subra, activist grandmother. *Guardian.co.uk,* June 20. http://www.guardian.co.uk/environment/2010/jun/20/tony-hayward-bp-oil-spill (accessed June 20, 2010).

Gonzalez, J. 2002. *Fallout: The environmental consequences of the World Trade Center collapse.* New York: The New Press.

Green, M. 1993. Interview with author. Lerwick, Shetland Islands, February 9.

Grossman, E. 2010a. All the data shows no toxic air concentrations—but response workers are stricken. *Pump Handle,* May 28. http://scienceblogs.com/thepumphandle/2010/05/all_the_data_shows_no_toxic_ai.php (accessed May 28, 2010).

———. 2010b. As we rush to protect the Gulf Coast environment, are responders being protected? *Pump Handle,* May 14. http://thepumphandle.wordpress.com/2010/05/14/are-gulf-coast-responders-being-protected/ (accessed May 14, 2010).

References

Guillemin, J. 1999. *Anthrax: The investigation of a deadly outbreak*. Berkeley: University of California Press.

Gusfield, J. 1981. *The culture of public problems*. Chicago: The University of Chicago.

Hadden, S. G. 1994. Citizen participation in environmental policy making. In *Learning from disaster*, edited by S. Jasanoff, 91–112. Philadelphia: University of Pennsylvania Press.

Haden, C. 2009. Perry County, Alabama takes on tons of toxic TVA sludge. *Birmingham Weekly*, July 9. http://www.bhamweekly.com/2009/07/09/perry-county-alabama-takes-on-tons-of-toxic-tva-sludge/ (accessed July 14, 2009).

Hagey, K. 2007. Dishonorable non-mention: Juan Gonzalez and the *Daily News'* 9/11 Pulitzer. *Village Voice: NYC Life*, April 17. http://www.villagevoice.com/2007-04-17/nyc-life/dishonorable-non-mention-juan-gonzalez-and-the-daily-news-9-11-pulitzer/ (accessed April 17, 2007).

Hall, M., 1993. Interview with author. Lerwick, Shetland Islands, February 9.

———. 2008. Probe: FEMA misdirected $13M in disaster aid. *USA Today*, February 21. http://www.usatoday.com/news/washington/2008-02-20-FEMA_N.htm (accessed February 23, 2008).

Hamlin, G. 2010. Interview with author. Conducted by phone. June 13.

Hammer, D. 2010. Advisers cited by Salazar say drilling ban is bad idea: Consultants sign letter disavowing six-month ban. *Times-Picayune*, June 9. http://www.nola.com/news/t-p/frontpage/index.ssf?/base/news-14/1276064428189870.xml&coll=1 (accessed June 9, 2010).

Hansen, J. 2009. *Storms of my grandchildren: The truth about the coming climate catastrophe and our last chance to save humanity*. New York: Bloomsbury.

Harding, S. 1991. *Whose science? Whose knowledge? Thinking from women's lives*. Ithaca, NY: Cornell University Press.

Harr, J. 1995. *A civil action*. New York: Vintage Books.

Harris, R. 2010. Gulf spill may far exceed official estimates. *National Public Radio*, May 14. http://www.npr.org/templates/story/story.php?storyId=126809525 (accessed May 14, 2010).

Hart, P. 2010. Still drill, baby—despite spill: Little rethinking of oil after Deepwater disaster. *Extra! Magazine of FAIR—The Media Watch Group* 23, no. 7 (July).

Harvey, D. 2005. *A brief history of neoliberalism*. New York: Oxford University Press.

———. 2006. *Spaces of global capitalism: Towards a theory of uneven geographical development*. London and New York: Verso.

Hayes, C. 2010. BP: Beyond punishment. *Nation*, June 9. http://www.thenation.com/article/bp-beyond-punishment (accessed June 11, 2010).

Health Task Force Newsletter. March, 1990.

Hearn, K. 2009. Tennessee's dirty data. *Nation*, April 2. http://www.thenation.com/article/tennessees-dirty-data?page=full (accessed September 13, 2009).

References

Hebert, H. J. 2009. EPA targets 44 coal ash sites in 10 states. *Knox News Sentinel*, June 30. http://www.knoxnews.com/news/2009/jun/30/epa-targets-44-coal-ash-sites-10-states/ (accessed June 30, 2009).

Hedrick. 1990. Interview with author. Homer, Alaska, April 18.

Heller, D. 2010. Florida scientist calls federal report on oil spill misleading. *WTSP.com*, August 20. http://www.wtsp.com/news/oilspill/stry.aspx?storyid=141638&catid=261 (accessed August 20, 2010).

Hennelly, R. 1990. Split wide open: Did the Valdez spill 11 million gallons—or 27 million? *Village Voice*, January 2.

Hennessy-Fiske, M., and R. Fausset. 2010a. Barbour blasts BP, federal government for cleanup effort. *Los Angeles Times*, June 28. www.latimes.com. (accessed June 28, 2010).

———. 2010b. Mississippi officials blast BP, U.S. government as oil hits coast. *Los Angeles Times*, June 29. http://www.latimes.com/news/nationworld/nation/la-na-oil-spill-20100629,0,1497067.story (accessed June 29, 2010).

Herbert, R. et al. 2006. The World Trade Center disaster and the health of workers: Five-year assessment of a unique medical screening program. *Environmental Health Perspectives* 114, no. 12: 1853–58.

Hilzenrath, D. S., and K. Kindy. 2010. Firms in Gulf drilling are working to limit liability in spill. *CommonDreams.org*, June 24. http://www.commondreams.org/headline/2010/06/24-9 (accessed June 24, 2010).

Hodgkinson, P. 1989. Technological disaster—Survival and bereavement. *Social Science and Medicine* 29, no. 3: 351–56.

Hoffman, A. J. 2001. *From heresy to dogma: An institutional history of corporate environmentalism*. Stanford, CA: Stanford University Press.

Hollis, J. 2004. Interview with author. Austin, Texas, March 15.

Homer Fisherman. 1990. Interview with author. Homer, Alaska, June 7.

Homer Layperson, 1990. Interview with author. Homer, Alaska, June 8.

Homer Multi-Agency Coordinating Group. 1989. Meeting Minutes April to August.

Homer Worker. 1990. Interview with author. Homer, Alaska, June 12.

Homer-area Resident. 1990. Interview with author. Homer, Alaska, June 12.

Hopkins, K. 2010. Health of *Exxon Valdez* cleanup workers was never studied. *McClatchy Washington Bureau*, June 29. http://www.mcclatchydc.com/2010/06/29/96782/health-of-exxon-valdez-cleanup.html#storylink=misearch (accessed June 29, 2010).

Hsu, S. S. 2008a. CDC confirms health risks to occupants of trailers. *Washington Post*, February 14. http://www.washingtonpost.com/wp-dyn/content/article/2008/02/13/AR2008021303937) (accessed February 17, 2008).

———. 2008b. Safety lapses raised risks in Katrina trailers. *MSNBC.com*, May 25. http://www.msnbc.msn.com/id/24810920 (accessed May 26, 2008).

References

———. 2010c. Deepwater horizon inspections: MMS skipped monthly inspections on doomed rig. *Huffington Post*, May 16,. http://www.huffingtonpost.com/2010/05/16/deepwater-horizon-inspect_n_578079.html (accessed May 16, 2010).

Huffington Post. 2010a. BP, Coast Guard officers block journalists from filming oil-covered beach. *Huffington Post*, May 19. http://www.huffingtonpost.com/2010/05/19/bp-coast-guard-officers-b_n_581779.html (accessed May 21, 2010).

———. 2010b. BP oil spill nears record as largest in Gulf history. *Huffington Post*, July 1. http://www.huffingtonpost.com/2010/07/01/bp-oil-spill-nears-record_n_631955.html (accessed July 1, 2010).

———. 2010c. Deepwater horizon inspections: MMS skipped monthly inspections on doomed rig. *Huffington Post*, May 16.

———. 2010d. Former EPA chief on Gulf oil spill: "It's going to blow the record books up." *Huffington Post*, June 11. http://www.huffingtonpost.com/2010/06/11/gulf-oil-spill-record-epa_n_609863.html (accessed June 12, 2010).

———. 2010e. Furious, Obama explains oil spill frustration to Larry King (VIDEO): "'Venting' won't solve anything," says Obama. *Huffington Post*, June 4. http://www.huffingtonpost.com/2010/06/04/furious-obama-explains-oi_n_600308.html (accessed June 4, 2010).

———. 2010f. Moratorium on deepwater drilling expected to be announced at Obama press conference. *Huffington Post*, May 27. http://www.huffingtonpost.com/2010/05/27/moratorium-on-deepwater-d_n_591560.html (accessed May 27, 2010).

Hvistendahl, M. 2007. Coal ash is more radioactive than nuclear waste: By burning away all the pesky carbon and other impurities, coal power plants produce heaps of radiation. *Scientific American*, December 13. http://www.scientificamerican.com/article.cfm?id=coal-ash-is-more-radioactive-than-nuclear-waste (accessed September 16, 2010).

Impact Assessment, Inc. 1990. *Final report: Economic, social, and psychological impact of the "Exxon Valdez" oil spill*. November 15. Anchorage, Alaska.

Inside Higher Education. 2010. Oil debate spills into Academe. *Inside Higher Ed.com*, July 20. www.insidehighered.com/views/2010/07/22/nelson (accessed July 20, 2010).

Isikoff, M., and M. Hirsh. 2010. Slick operator: How British oil giant BP used all the political muscle money can buy to fend off regulators and influence investigations into corporate neglect." *Newsweek*, May 7. http://www.newsweek.com/2010/05/07/slick-operator.html (accessed May 7, 2010).

Iyengar, S. 1991. *Is anyone responsible? How television frames political issues*. Chicago: University of Chicago Press.

Jackson, D. 2010a. Big oil & conflicts of interest: Obama owns this mess. *CommonDreams.org*, May 22. http://www.commondreams.org/view/2010/05/22-7 (accessed May 22, 2010).

References

———. 2010b. Troubled chickens come home to roost. *Boston Globe,* May 22. http://www.boston.com/bostonglobe/editorial_opinion/oped/articles/2010/05/22/troubled_chickens_come_home_to_roost/ (accessed May 22, 2010).

January 28 Committee. n.d. *The Santa Barbara declaration of environmental rights.* Santa Barbara, CA: Noel Young Press.

Jarvie, J. 2008. Displaced by Katrina and edged out of FEMA trailer parks. *Common-Dreams.org,* June 1. http://www.commondreams.org/archive/2008/06/01/9344 (accessed June 1, 2008).

Jasanoff, S. 1988. The Bhopal disaster and the right to know. *Social Science and Medicine 27,* no. 10: 1113–23.

Jervis, R. 2009. Report: FEMA mishandled toxins in trailers. *CommonDreams. org,* July 24. http://www/commondreams.org/print/45063 (accessed July 24, 2009).

Johnson, V. 2006. Interview with author, April 14.

Johnston, B. R., and H. M. Barker. 2008. *Consequential damages of nuclear war.* Walnut Creek, CA: Left Coast Press.

Jones, S. 2010. The media-lobbying complex. *Nation,* March 1. http://www.thenation.com/article/media-lobbying-complex?page=0,2. (accessed March 1, 2010).

Juhasz, A. 2008. *The tyranny of oil: The world's most powerful industry—and what we must do to stop it.* New York: William Morrow.

Kaufman, L., and J. McKinley, Jr. 2010. Cleanup draws critics over speed and care. *New York Times,* May 30. http://www.nytimes.com/2010/05/31/us/31cleanup. html?scp=1&sq=%22Cleanup%20draws%20critics%20over%20speed%20 and%20care%22&st=cse (accessed May 30, 2010).

Kearnes, S. n.d. Poisoned waters: Alaska natives and the oil spill. Independently Produced Radio Documentary. Homer, Alaska: National Public Radio.

Keating, M., L. Evans, B. Dunham, and J. Stant. n.d. Waste deep: Filling mines with coal ash is profit for industry, but poison for people. *Earth Justice,* January 15. http://www.earthjustice.org/library/reports/earthjustice_waste_deep. pdf (accessed January 15, 2009).

Keeble, J. 1991. *Out of the channel: The "Exxon Valdez" oil spill in Prince William Sound.* New York: Harper Collins.

Keesing, R. 1987. Anthropology as interpretive quest. *Current Anthropology 28:* 161–76.

Keil, A. 2009. Roane residents worry about arsenic test results. *6WATE.com,* January 2. http://www.wate.com/Global/story.asp?S=9610602 (accessed January 10, 2009).

Kendall, M. L. 2010. Memorandum to K. Salazar. Investigative report—Island operating company, et al. May 24. http://www.eenews.net/public/25/15844/features/documents/2010/05/25/document_gw_02.pdf.

Kennedy, R. F., Jr. 2010. Sex, lies and oil spills. *Huffington Post,* May 5. http://www.huffingtonpost.com/robert-f-kennedy-jr/sex-lies-and-oil-spills_b_564163.html (accessed May 5, 2010).

References

Kilcher, J. 1990. Interview with author, August 10.

Kincaid, L. 2009. Formaldehyde Council, Inc. disputes data on formaldehyde in homes. *Examiner*, July 21. http://www.Examiner.com/x-5101-San-Jose-Environmental-Health-Examiner (accessed July 26, 2009).

Kingston Resident. 2009. Interview with author. Roane County, Tennessee, February.

Kirkham, C. 2010a. Chalmette community meeting on Gulf of Mexico oil spill unleashes anger, frustration. *Times-Picayune*, May 24. http://www.nola.com/news/gulf-oil-spill/index.ssf/2010/05/chalmette_community_meeting_on.html (accessed May 25, 2010).

———. 2010b. Coast Guard puts up 65-foot cleanup buffer: News media, public can't get close. *Times-Picayune*, July 2. http://www.nola.com/news/t-p/frontpage/index.ssf?/base/news-15/127805231036310.xml&coll=1 (accessed July 2, 2010).

Kizzia, T. 1989. Residents gain more control. *Homer News*, April 27, 1 and 24.

Klein, N. 2007. *The shock doctrine: The rise of disaster capitalism.* New York: Metropolitan Books.

Knight, F. H. 1921. *Risk, uncertainty and profit.* New York: Harper and Row.

Knoxville News Sentinel. 2008a. Neighbor Chris Copeland after the TVA pond breach. December 23. http://www.knoxnews.com/videos/detail/neighbor-chris-copeland-after-tva-pond-breach/ (accessed December 23, 2008).

———. 2008b. Mudslide from TVA pond breach closes Emory River. December 23, 2008. http://www.knoxnews.com/news/2008/dec/23/mudslide-tva-pond-breach-closes-emory-river/?partner=RSS (accessed December 23, 2008).

Knudson, T. 2010. Quest for oil leaves trail of damage across the globe. *McClatchy*, May 16. http://www.mcclatchydc.com/2010/05/16/94126/quest-for-oil-leaves-trail-of.html (accessed May 16, 2010).

Kompkoff, G. 1990. Interview with author. Homer, Alaska, April 18.

Krell, R., and L. Rabkin. 1979. The effects of sibling death on the surviving child: A family perspective. *Family Process* 18: 471–77.

Krimsky, S. 2000. *Hormonal chaos: The scientific and social origins of the environmental endocrine hypothesis.* Baltimore: The Johns Hopkins University Press.

———. 2003. *Science in the private interest: Has the lure of profits corrupted biomedical research?* Lanham, MD: Rowman and Littlefield Publishers.

———. 2007. Exxon funds litigation research, gets reduced damages. Pump Handle. January 16. http://thepumphandle.wordpress.com/2007/01/16/exxon-funds-litigation-research-gets-reduced-damages/ (accessed January 16, 2007).

Krogman, N. 1996. Frame disputes in environmental controversies: The case of wetland regulations in Louisiana. *Sociological Spectrum* 16:371–400.

Kroll-Smith, S., and H. H. Floyd. 1997. *Bodies in protest: Environmental illness and the struggle over medical knowledge.* New York: New York University Press.

Kroll-Smith, S., P. Brown, and V. J. Gunter., eds. 2000. *Illness and the environment: A reader in contested medicine.* New York: New York University Press.

References

Kunzelman, M. 2008. FEMA asks federal judge for immunity from Katrina victims' lawsuits over trailer fumes. *Minneapolis Star Tribune*, July 23. http://www.star-tribune.com/ (accessed July 24, 2008).

Kvasnikoff, J. 1990. Interview with author. Homer, Alaska, April 23.

Lapierre, D., and J. Moro. 2002. *Five past midnight in Bhopal: The epic story of the world's deadliest industrial disaster.* New York: Warner Books.

Latin, L. 1988. Good science, bad regulation, and toxic risk assessment. *Yale Journal on Regulation* 5: 89–148.

Lave, L. 1986. Approaches to risk management: A critique. In *Risk Evaluation and Management*, edited by V. T. Covello, J. Menkes, and J. Mumpower, 93–130. New York: Plenum Publishing Corporation.

Leland, J. 2010. Cleanup hiring feeds frustration in fishing town. *New York Times*, June 26. http://www.nytimes.com/2010/06/27/us/27bayou.html?scp=1&sq=Cleanup%20Hiring%20Feeds%20Frustration%20in%20Fishing%20Town&st=cse (accessed June 28, 2010).

Levin, A. 2010. Oil spills escalated in this decade. *USA Today*, June 8. http://www.usatoday.com/news/nation/2010-06-07-oil-spill-mess_N.htm (accessed June 8, 2010).

Levine, A. 1982. *Love Canal: Science, politics, and people.* Lexington, MD: Lexington Books.

Levine, B. 1989. Letter to the Editor. *Homer (Alaska) News*, July 27, 6.

Lewis, J. 2006. Interview with author. St. Bernard Parish, Louisiana, March 18.

Lewis, W. 2009. Beasley Allen files coal ash spill class action lawsuit on behalf of residents and property owners affected. Beasley, Allen, Crow, Methvin, Portis & Mile, P.C. January 9. http://www.coal-ash-spill.com/news/2009/01/09/beasley-allen-files-coal-ash-spill-class-action-lawsuit-on-behalf-of-residents-and-property-owners-affected (accessed July 15, 2009).

Lifton, R. J., and E. Olson. 1976. The human meaning of a total disaster: The Buffalo Creek experience. *Psychiatry* 39: 1–18.

Linds, R., and A. Linds. 1990. Interview with author. August 18, 1990.

Linkins, J. 2010a. BP better at stemming journalists than oil wells. *Huffington Post*, May 25. http://www.huffingtonpost.com/2010/05/25/bp-better-at-stemming-jou_n_589260.html (accessed May 27, 2010).

———. 2010b. BP media clampdown: Journalists now face possibility of fines, prison time. *Huffington Post*, July 6. http://www.huffingtonpost.com/2010/07/06/bp-media-clampdownjourna_n_636317.html?page=29&show_comment_id=52705741#comment_52705741 (accessed July 6, 2010).

———. 2010c. BP sends PR professionals to Gulf Coast to pretend to be journalists. *Huffington Post*, June 27. http://www.huffingtonpost.com/2010/06/24/bp-sends-pr-professionals_n_624686.html (accessed June 27, 2010).

———. 2010d. Oil spill lawsuits: BP spending big to acquire an army of expert witnesses. *Huffington Post*, July 16. http://www.huffingtonpost.com/2010/07/16/oil-spill-lawsuits-bp-spe_n_649335.html (accessed July 16, 2010).

———. 2010e. Oil spill response: "Army of temp workers" bused to Grand Isle for Obama appearance leave soon afterward. *Huffington Post*, May 29. http://www.huffingtonpost.com/2010/05/28/oil-spill-response-army-o_n_594014.html (accessed May 29, 2010).

Liptak, A. 2008. "From one footnote, a debate over the tangles of law, science, and money. *New York Times*, November 24. http://www.nytimes.com/2008/11/25/washington/25bar.html?_r=1 (accessed November 24, 2008).

Lodge, B. 2009. Storm-related lawsuits involving trailers jump. *2theadvocate.com*, July 21. http://2theadvocate.com/news/51271747.html?showAll=y&c=y (accessed July 26, 2009).

Losbaugh, S. 1990. Interview with author. Homer, Alaska, August 3.

Louisiana Bucket Brigade. 2006. "Science for sale: How Murphy Oil, USA and the Center for Toxicology and Environmental Health have downplayed health risks from Murphy's million gallon oil spill in St. Bernard Parish."

Louisiana Bucket Brigade Community Meeting. 2006. Chalmette, Louisiana, March 19.

Lubchenco, J. 2009. Letter to S. E. Birnbaum. September 21, 2009. http://www.peer.org/docs/noaa/09_12_10_NOAA_Comments_on_MMS_5_Year_Plan.pdf.

Lucas, T. 2009. Toxic coal ash threatens health and environment: Kingston TVA spill could create long-term exposures to area residents. *Office of News and Communications, Duke University*, August 18. http://news.duke.edu/2009/08/toxiccoal.html (accessed August 21, 2009).

Lustgarten, A., and R. Knutson. 2010. Years of internal BP probes warned that neglect could lead to accidents. *ProPublica.org*, June 7. http://www.propublica.org/feature/years-of-internal-bp-probes-warned-that-neglect-could-lead-to-accidents (accessed June 7, 2010).

Lyons, R. 1999. Acute health effects of the Sea Empress oil spill. *Journal of Epidemiology and Community Health*. May, 53 (5): 306–310.

M. H. 1991. When science is sealed by the courts. *Scientific American* 265.4.

Mackin, B. 2006. Interview with author. St. Bernard Parish, Louisiana, February 27.

Maddow Show, The. 2010. May 26. http://www.msnbc.msn.com/id/37376039/ns/msnbc_tv-rachel_maddow_show/ (accessed May 26, 2010).

Mansfield, D. 2009a. Engineer: TVA ash spill resulted from dike burst. *Knox News Sentinel*, July 10. http://www.knoxnews.com/news/2009/jul/10/engineer-tva-ash-spill-resulted-dike-burst/ (accessed July 14, 2009).

———. 2009b. Memo details TVA editing of response to ash spill. *Knox News Sentinel*, January 23. http://www.knoxnews.com/news/2009/jan/23/memo-details-tva-editing-response-ash-spill/ (accessed January 26, 2009).

References

———. 2009c. TVA ash spill site fails to make EPA hazard list. *Knox News Sentinel*, July 13. http://www.knoxnews.com/news/2009/jul/13/tva-ash-spill-site-fails-make-epa-hazard-list/ (accessed July 14, 2009).

Marcum, E. 2009a. Analysis of ash release due Oct. 10. *Knox News Sentinel*, October 2. http://www.knoxnews.com/news/2009/oct/02/analysis-of-ash-release-due-oct-10/ (accessed on October 3, 2009).

———. 2009b. Residents affected by coal ash spill critical of TVA: Many say incident ruined their property but claims rejected. *Knox News Sentinel*, June 24. http://www.knoxnews.com/2009/jun/24/residents-critial-of-tva (accessed June 25, 2009).

Markowitz, G., and D. Rosner. 2002. *Deceit and denial: The deadly politics of industrial pollution*. Berkeley: University of California Press.

Marquardt, K. 1990. Interview with author. August 11, 1990.

Marrett, C. 1981. The accident at Three Mile Island and the problem of uncertainty. In *The Three Mile Island nuclear accident: Lessons and implications*, edited by D. Stills and T. Moss, 280–91. Annals of the New York Academy of Sciences 365. New York: New York Academy of Sciences.

Martin, E. Letter to John G. Morgan, June 5, 2009.

Mattei, U., and L. Nader. 2008. *Plunder: When the rule of law is Illegal*. Malden, MA: Blackwell Publishing.

Maugh, T. H., and J. Jarvie. 2008. FEMA trailers toxic, tests show: Unhealthy levels of formaldehyde are found. The agency will expedite efforts to relocate occupants. *Los Angeles Times*, February 15. http://www.latimes.com/news/nationworld/nation/la-na-trailers15feb15,1,1726106.story (accessed February 17, 2008).

Mazur, A. 1981. *The dynamics of technical controversy*. Washington, DC: Communications Press.

McCarthy, M. 2009. TVA prepares to move spilled fly ash from Roane County. Channel 8, WVLT, June 23. http://www.volunteer.tv.com (accessed July 14, 2009).

McChesney, R. W., and J. Nichols. 2010. *The death and life of American journalism: The media revolution that will begin the world again*. Philadelphia: Nation Books.

McCright, A., and R. Dunlap. 2003. Defeating Kyoto: The conservative movement's impact on U. S. climate change policy." *Social Problems*. Vol. 50, No. 3: 348–373.

McClelland, M. 2010. It's BP's oil: Running the corporate blockade at Louisiana's crude-covered beaches. *Mother Jones*, May 24. http://motherjones.com/environment/2010/05/oil-spill-bp-grand-isle-beach (accessed May 24, 2010).

McDaniel, M. n.d. Letter to the Editor. *Department of Environmental Quality, Louisiana*. http://www.deq.state.la.us/portal/portals/0/news/pdf/McDaniel-LettertotheEditor123005.pdf.

McGarity, T. O., and W. E. Wagner. 2008. *Bending science: How special interests corrupt public health research*. Cambridge, MA: Harvard University Press.

McMullen, E. 1990. Interview with author. April 20, 1990.

McNulty, P. 1990. Exxon's problem: Not what you think. *CNN.com*, April 23.

Merry, S. 1990. *Getting justice and getting even*. Chicago: University of Chicago Press.

Michaels, D. 2008. *Doubt is their product: How industry's assault on science threatens your Health*. New York: Oxford University Press.

Milam, G. 2010. BP chief: Oil spill impact "very modest." *Sky News*, May 18. http://news.sky.com/skynews/Home/World-News/BP-Oil-Spill-In-Gulf-Of-Mexico-Will-Have-Very-Modest-Environmental-Impact-Says-Firms-CEO/Article/20 1005315633987?lpos=World_News_Top_Stories_Header_2 (accessed May 18, 2010).

Mileti, D., D. Drabeck, and J. Hass. 1975. *Human systems in extreme environments*. Boulder: University of Colorado Institute of Behavioral Science.

Miner, E. 2004. Fighting pollution in St. Bernard Parish. *Louisiana Environmental Justice Voices*. Vol. 1, Issue 3. November.

Molotch, H. 1972. Oil in Santa Barbara and power in America. *Sociological Inquiry* 4: 131–44.

Monbiot, G. 2010. The oil firms' profits ignore the real costs: The energy industry has long dumped its damage and, like the banks, made scant provision against disaster. Time to pay up. *CommonDreams.org*, June 8. http://www.commondreams.org/view/2010/06/08 (accessed June 8, 2010).

Moore, S. F., and B. G. Myerhoff, eds. 1975. *Symbol and politics in communal ideology: Cases and questions*. Ithaca, NY, and London: Cornell University Press.

Morano, M. 2003. Alarmist global warming claims unfounded says climatologist. *Free Republic*, July 14. http://www.freerepublic.com/focus/f-news/945664/posts (accessed July 14, 2003).

Moreno Gonzales, J. 2008. Kids in Katrina trailers may face lifelong ailments. *Associated Press*, May 28. http://ap.google.com/article/ALeqM5i0obdVKtwUqMG-6gEUqc6mnjat9hAD90UESS00 (accessed May 28, 2008).

Morgan, D. 2010. Adm. Allen: No need for Gov't to take over spill. CBS News. May 24. http://www.cbsnews.com/stories/2010/05/24/national/main6514829.shtml (accessed May 24, 2010).

Morrison, C. 2009. NC: No delay in TVA's cleanup. *Asheville Citizen Times*, July 12. http://www.citizen-times.com/ (accessed July 14, 2009).

Morton, J. 2009. An estimated 39 million tons of coal ash destined for Perry County landfill. *Tuscaloosa News*, June 28. http://www.tuscaloosanews.com/article/20090628/NEWS/906279948 (accessed June 29, 2009).

Mouawad, J. 2010. For BP, a history of spills and safety lapses. *New York Times*, May 8. http://www.nytimes.com/2010/05/09/business/09bp.html (accessed May 8, 2010).

Mouawad, J., and C. Krauss. 2010. Another torrent BP works to stem: Its C.E.O. *New York Times*, June 3. http://www.nytimes.com/2010/06/04/us/04image. html?scp=1&sq=%22Another%20torrent%20BP%20works%20to%20stem:%20 Its%20C.E.O%22&st=cse (accessed June 3, 2010).

Mowbray, Rebecca. 2010. Timor Sea blowout foreshadowed disaster in Gulf. *Times-Picayune*, June 13. http://www.nola.com/news/t-p/frontpage/index.ssf?/base/ news-14/1276410019283560.xml&coll=1 (accessed June 13, 2010).

MSNBC.com. 2007. FEMA trailers at fire-sale prices: Mobile-home dealers fear government will flood the market. *MSNBC.com*, March 8. http://www.msnbc.com/ id/17509045 (accessed March 8, 2009).

Mufson, S. 2010. Federal records show steady stream of oil spills in gulf since 1964. *Washington Post*, July 24. www.washingtonpost.com (accessed July 24, 2010).

Mulkern, A. 2008. Sex, drugs alleged in oil deals. *Denver Post*, September 11. http://www.denverpost.com/news/ci_10431998/http://www.nytimes.com/ gwire/2010/05/25/25greenwire-interior-probe-finds-fraternizing-porn-and-dru-45260.html/ (accessed September 11, 2008).

Munger, F. 2009a. Coal ash spill-related research program soliciting proposals. *Knox News Sentinel*, July 14. http://www.knoxnews.com/news/2009/jul/14/research-program-soliciting-proposals/ (accessed July 14, 2009).

———. Frank Munger's Atomic City Underground. http://blogs.knoxnews.com/ munger/2009/08/oraus_med_evaluations_for_fly-.html. (accessed September 1, 2009).

Murky waters. 2010. *New York Times*, May 20. http://www.nytimes.com/2010/05/21/ opinion/21fri1.html (accessed May 20, 2010)

Murphy, K. 2001. Exxon spill's cleanup workers share years of crippling illness. *Los Angeles Times*. November 5.

Mysuburblife.com. 2009. TVA: Letter from Tom Kilgore, TVA CEO. Blog. January 15. http://www.mysuburblife.com/2009/01/tva-letter-from-tom-kilgore-tva-ceo.html (accessed January 15, 2009).

Nash, R. 1985. *Wilderness and the American mind*. New Haven, CT: Yale University Press.

———. 1989. *The rights of nature: A history of environmental ethics*. Madison: The University of Wisconsin Press.

National Institute for Occupational Safety and Health (NIOSH). 1991. *Exxon-Valdez* Alaska oil spill. Health Hazard Evaluation Report. HETA 89–200, 89–273–2111. May.

National Research Council. 1989. *Using oil dispersants on the sea*. Washington, DC: National Academy Press.

National Response Team. 1989. *The "Exxon Valdez" oil spill: A report to the president*. Washington, DC: U.S. Department of Transportation.

References

National Wildlife Federation Commission. 1989. *The day winter died. A compilation of the November 1989 Citizen's Commission hearings on the "Exxon Valdez" oil spill.* Anchorage, Alaska.

Natural Resource Defense Council. 1990. *No safe harbor: Tanker safety in America's ports.* New York: Natural Resource Defense Council.

———. 2009. NRDC urges immediate clean up and stronger regulations of coal waste: Response to senate hearing on Tennessee sludge spill. Press release, January 8. http://www.nrdc.org/media/2009/090108.asp (accessed January 10, 2009).

Naureckas, J. 2010. Managed news from the Gulf of Mexico. *Fairness and accuracy in reporting,* May 27. http://www.fair.org/blog/2010/05/27/managed-news-from-the-gulf-of-mexico/ (accessed May 31, 2010).

Navis, J. 2006. Interview with author. Austin, Texas, February 26.

Navis, S. 2006. Interview with author. Austin, Texas, February 26.

Naylor, B. 2009. FEMA works to avoid formaldehyde in new units. *National Public Radio,* May 18. http://www.npr.org/templates/story/story.php?storyId=104203598 (accessed May 18, 2009).

Nelkin, D., and M. S. Brown. 1982. *Workers at risk: Voices from the workplace.* Chicago: University of Chicago Press.

Nelkin, D., ed. 1985. *The language of risk: Conflicting perspectives on occupational health.* Beverly Hills, CA: Sage Publications.

Nelson, C. 2010. BP and academic freedom. *Inside Higher Education,* July 22. http://www.insidehighered.com/views/2010/07/22/nelson (accessed July 25, 2010).

NewsInferno.com. 2008a. FEMA gave little thought to safety of Katrina trailers. *NewsInferno.com,* May 27. http://www.newsinferno.com/archives/3151 (accessed May 28, 2008).

———. 2008b. FEMA trailer resident details health problems for lawmakers. *NewsInferno.com,* April 3. http://www.newsinferno.com/archives/2841 (accessed April 12, 2008).

———. 2009. First toxic FEMA trailer lawsuit scheduled for September trial. *NewsInferno.com,* April 8. http://www.newsinferno.com/archives/5516 (accessed July 26, 2009).

NOAA Official. 1990. Interview with author. Homer, Alaska, April 10.

Noguchi, Y. 2010. Cleanup jobs are hard to find in the Gulf. *National Public Radio,* June 28. http://www.npr.org/templates/story/story.php?storyId=128109282 (accessed June 28, 2010).

Norman, F. 1990. Interview with author, April 20, 1990.

Norris, M. 2006. Letter to S. Abadi, November 29. http://www.chinesedrywall.com/files/CTECH_Report.pdf (accessed July 1, 2010).

Norton, M. 1993. Interview with author. Lerwick, Shetland Islands, February 13.

References

Nuckols, B. 2010. Gulf oil spill illness: Four hospitalized after getting sick, 125 cleanup boats recalled. *Huffington Post*, May 27. http://www.huffingtonpost.com/2010/05/27/gulf-oil-spill-illness-st_n_591510.html (accessed May 27, 2010).

O'Carroll, E. 2008. Manufacturers say they knew of FEMA trailer health risks. *Christian Science Monitor*, July 11. http://features.csmonitor.com/environment/2008/07/11/manufacturers-say-they-knew-of-fema-trailer-health-risks/ (accessed July 26, 2009).

Occupational Health and Safety. 2008. CDC, FEMA: Formaldehyde levels in Gulf Coast trailers too high. *Occupational Health & Safety*, February 15. http://www.ohsonline.com/ (accessed February 17, 2008).

Oil and Gas Journal. 1990. Double hulls, officer training, key tanker safety issues. *Oil and Gas Journal* 18 (June): 15.

Oldham, R. 1990. Interview with author. Homer, Alaska, August 17.

Oliver-Smith, A. 2004. Theorizing vulnerability in a globalized world: A political ecological perspective. In *Mapping vulnerability: Disasters, development & people,* edited by G. Bankoff, G. Frerks, and D. Hilhorst. London: Earthscan.

Oliver-Smith, A., and S. M. Hoffman. 1999. *The angry earth: Disaster in anthropological perspective.* New York: Routledge.

———. 1990. Interview with author. Homer, Alaska, June 14.

O'Meara, J. 1989. *Cries from the heart. Alaskans respond to the "Exxon Valdez" oil spill.* Homer, AK: Wizard Works Press.

———. 1990. Interview with author. Homer, Alaska, June 14.

On the Media. 2010. How much oil really spilled from the *Exxon Valdez*? National Public Radio's *On the Media*, June 18. http://www.onthemedia.org/transcripts/2010/06/18/01 (accessed June 18, 2010).

Online Magazine of the Institute for Southern Studies. n.d. Coal ash contamination imperils July 4 festival goers in Tennessee. Institute for Southern Studies. http://www.southernstudies.org/2009/07/coal-ash-contamination-imperils-july-4-festival-goers-in-tennessee.html (accessed July 6, 2009).

———. n.d. Decision to dump TVA's spilled coal waste in Alabama community sparks resistance. Institute for Southern Studies. http://www.southernstudies.org/2009/07/decision-to-dump-tvas-spilled-coal-waste-in-alabama-community-sparks-resistance.html (accessed July 14, 2009).

———. n.d. Katrina trailer contractor failed to act on known health risks. Institute for Southern Studies. http://www.southernstudies.org/2008/07/katrina-trailer-contractor-failed-to.html (accessed June 21, 2009).

———. n.d. Was the Tennessee coal ash disaster really a once-in-a-lifetime event? Institute for Southern Studies. http://www.southernstudies.org/2009/06/was-the-tennessee-coal-ash-disaster-really-a-once-in-a-lifetime-event.html (accessed June 30, 2009).

Oreskes, N., and E. M. Conway. 2010. *Merchants of doubt: How a handful of scientists obscured the truth on issues from tobacco smoke to global warming.* New York: Bloomsbury Press.

Ortega, B. 1989a. Coast Guard overruled other groups: Exxon left despite complaints. *Homer News*, September 21, 1 and 36.

———. 1989b. Otter activists battle U.S. Fish and Wildlife. *Homer News*, September 7, 17 and 40.

Ortner, S. B. 2006. *Anthropology and social theory: Culture, power, and the acting subject.* Durham, NC: Duke University Press.

Ott, R. 2005. *Sound truth and corporate myths: The legacy of the "Exxon Valdez" oil spill.* Cordova, AK: Dragonfly Sisters Press.

———. 2010a. At what cost? BP spill responders told to forgo precautionary health measures in cleanup. *Huffington Post*, May 17. http://www.huffingtonpost.com/riki-ott/at-what-cost-bp-spill-res_b_578784.html (accessed May 18, 2010).

———. 2010b. From the ground: BP censoring media, destroying evidence. *Huffington Post*, June 11. http://www.huffingtonpost.com/riki-ott/from-the-ground-bp-censor_b_608724.html (accessed June 11, 2010).

Ozonoff, D., and L. Boden. 1987. Truth and consequences: Health agency responses to environmental health problems. *Science, Technology, and Human Values* 12, no. 3/4: 70–77.

Padgett, T. 2009. Is drywall the next Chinese import scandal? *Time*, March 23. http://www.time.com/time/nation/article/0,8599,1887059,00.html (accessed March 23, 2009).

Paine, A. 2009. Benton County neighbors fear dangers of TVA's coal ash: Health complaints investigated. *Tennessean*, June 20. http://www.tennessean.com/apps/pbcs.dll/article?AID=/20090620/NEWS0201/906200347 (accessed June 22, 2009).

Parker, Waichman, Alonso LLP. n.d. Toxic FEMA trailer lawsuit lawyers. http://www.yourlawyer.com/topics/overview/toxic_fema_trailers (accessed June 21, 2009).

Parman, S. 1990. *Scottish crofters. A historical ethnography of a Celtic village.* Fort Worth, TX: Holt, Rinehart and Winston.

Payne, D. 2006. Interview with author. St. Bernard Parish, Louisiana, August.

Peacock, A. 2003. *Libby, Montana: Asbestos and the deadly silence of an American corporation.* Boulder, CO: Johnson Books.

Perez-Cadahia, B., et al. 2008. Relationship between blood concentrations of heavy metals and cytogenic and endocrine parameters among subjects involved in cleaning coastal areas affected by the Prestige tanker oil spill. *Chemosphere* 71: 447–55.

Perez-Cadahia, B., B. Laffon, V. Valdiglesias, E. Pasaro, and J. Mendez. 2008. Cytogenetic effects induced by Prestige oil on human populations: The role of polymorphisms in genes involved in metabolism and DNA repair. *Mutation Research* 653: 117–23.

Perez-Cadahia, B., et al. 2008. Biomonitoring of human exposure to Prestige Oil: Effects on DNA and endocrine parameters. *Environmental Health Insights* 2: 83–92.

References

Perry, R., and M. Lindell. 1978. The psychological consequences of natural disaster: A review on American communities. *Mass Emergencies* 3: 105–15.

Peters, J. W. 2010. Efforts in Gulf to limit flow of spill news. *New York Times*, June 9. http://www.nytimes.com/2010/06/10/us/10access.html?fta=y (accessed June 9, 2010).

Plushnick-Masti, R., and N. Schwartz. 2010. Gulf oil spill: BP tries to block release of oil spill research. *Huffington Post*, June 24. http://www.huffingtonpost.com/2010/07/24/gulf-oil-spill-bp-tries-t_0_n_658300.html (July 25, 2010).

Potter, J. 1973. *Disaster by oil. Oil spills: Why they happen, what they do, how we can end them.* New York: Macmillan Company, reprinted courtesy of Marine Pollution Control/MPC Environmental Services, Calverton, New York.

Preston, J. 2006. Public misled on air quality after 9/11 attack, judge says. *New York Times*, February 3. http://www.nytimes.com/2006/02/03/nyregion/03suit.html?_r=1&scp=1&sq=February%203,%202006,%20Whitman&st=cse (accessed February 3, 2006).

Proctor, J. 2009. As MRGO nears closure, takings suit proceeds. *SeaGrant: Mississippi-Alabama Legal Program.* May. http://masglp.olemiss.edu/Water%20Log/WL29/29.1mrgo.htm. (accessed May 30, 2009).

Public Employees for Environmental Responsibility (PEER). 2010a. NOAA concerns brushed aside in Obama offshore drilling plan. Press release, April 1. http://www.peer.org/news/news_id.php?row_id=1324 (accessed April 1, 2010).

———. 2010b. Obama's orphaned science integrity and transparency pledge. Press release, July 8. http://www.peer.org/news/news_id.php?row_id=1371 (accessed July 8, 2010).

Pump Handle. 2010. Lessons from *Exxon Valdez* on worker health and safety. *Pump Handle,* May 3. http://thepumphandle.wordpress.com/2010/05/03/lessons-from-exxon-valdez-on-worker-health-and-safety/ (accessed May 3, 2010).

Raines, B. 2010. BP buys up Gulf scientists for legal defense, roiling academic community. *Alabama Local News,* July 16. http://blog.al.com/live/2010/07/bp_buys_up_gulf_scientists_for.html (accessed July 19, 2010).

Raloff, J. 2010. Feds probe Gulf spill health risks. *Science News,* June 17. http://www.sciencenews.org/view/generic/id/60373 (accessed June 17, 2010).

Rampton, S., and J. Stauber. 2001. *Trust us, we're experts: How industry manipulates science and gambles with your future.* New York: Tarcher/Putnam.

Raphael, B. 1986. *When disaster strikes.* New York: Basic Books.

———. 1991. Rescuer's psychological responses to disasters. *British Medical Journal* 303 (6814): 1346–47.

Raphael, B., B. Singh, L. Bradbury, and F. Lambert. 1983–1984. Who helps the helpers? The effects of a disaster on rescue workers. *Omega* 15, no. 1: 9–20.

Rappaport, R. 1993. The anthropology of trouble. *American Anthropologist* 95: 295–303.

References

Rappaport, R. A. 1988. Toward postmodern risk analysis. *Risk Analysis* 8, no. 2: 189–91.

Reed, Carol. 1985. The role of wild resources use in communities of the central Kenai Peninsula and Kachemak Bay. Technical Report Paper No. 106, October. Anchorage: Alaska Department of Fish and Game, Division of Subsistence.

Reeves, J. 2009. DA calls dumping ash in Uniontown "tragic." *Montgomery Advertiser*, July 8. http://www.montgomeryadvertiser.com/article/20090708 (accessed July 14, 2009).

Reich, M. 1991. *Toxic politics. Responding to chemical disasters.* Ithaca, NY: Cornell University Press.

Reich, R. 2010a. Why Obama should put BP under temporary receivership. *Huffington Post,* May 31. http://www.huffingtonpost.com/robert-reich/why-obama-should-put-bp-u_b_595346.html (accessed May 31, 2010).

———. 2010b. Why the United States still can't get BP to do what's necessary. *Huffington Post,* June 13. http://www.huffingtonpost.com/robert-reich/why-the-united-states-sti_b_610469.html (accessed June 13, 2010).

Reller, C. 1993. Occupational exposures from oil mist during the EVOS cleanup. Abstract. *EVOS Symposium Abstracts,* 313–315.

Rich, F. 2010a. Clean the Gulf, clean house, clean their clock. *New York Times,* June 18. http://www.nytimes.com/2010/06/20/opinion/20rich.html?scp=1&sq=Clean%20the%20Gulf,%20clean%20house,%20clean%20their%20clock&st=cse (accessed June 18, 2010).

———. 2010b. Don't get mad, Mr. President. Get even. *New York Times,* June 4. http://www.nytimes.com/2010/06/06/opinion/06rich.html?scp=1&sq=%22Don't%20Get%20Mad&st=cse (accessed June 4, 2010).

———. 2010c. Obama's Katrina? Maybe worse. *New York Times,* May 28. http://www.nytimes.com/2010/05/30/opinion/30rich.html?scp=1&sq=Obama's%20Katrina?%20Maybe%20worse&st=cse (accessed May 28, 2010).

Richardson, J. 2010. The "scathing internal memo" about BP's failures to protect worker safety. *Lavidalocavore.org,* June 3. http://www.lavidalocavore.org/showDiary.do?diaryId=3646&view=print (accessed on June 3, 2010).

Rioux, P. 2007. Katrina survivors fight Murphy Oil permit for more pollution and more tanks after disaster. *Global Community Monitor,* July 1. http://www.shell-facts.com/article.php?id=599 (accessed March 8, 2008).

Roane County Couple. 2009. Interview with author. Roane County, Tennessee, February 10, 2009.

Roane County Mother. 2009. Interview with author. Roane County, Tennessee, February, 2009.

Roane County Resident. 2009. Interview with author. Roane County, Tennessee, February, 2009.

Roane Views. 2009a. TVA/TDEC/EPA public meeting last night. *Roane Views,* October 2. http://www.roaneviews.com/node/3936 (accessed October 3, 2009).

References

———. 2009b. WUOT podcast with Dr. Gregory Button on TVA disaster. *Roane Views*. http://www.roaneviews.com/?q=node/2928 (accessed May 20, 2009).

Rolfes, A., K. Ford, and A. Babich. 2003. Citizen groups give Chalmette Refining notice of their intent to sue for violations of the Clean Air Act and the Emergency Planning and Community Right to Know Act. *Tulane Law School*. Press release, December 4. http://www.refineryreform.org/PR_Tulane_120403 (accessed January 17, 2007).

Rosen, D. H., and D. Voorhees-Rosen. 1978. The Shetland Islands: The effects of social and ecological change on mental health. *Culture, Medicine and Psychiatry* 2: 41–68.

Rosen, L. 1984. *Bargaining for reality*. Chicago: University of Chicago Press.

Rowlands, C. 1993. Interview with author. Lerwick, Shetland Islands, February 15.

Russell, S., S. Lewis, and B. Keating. 1992. Inconclusive by design: Waste, fraud and abuse in federal environmental health research. *Environmental Health Network National Toxics Campaign Fund*, May.

Ryan, C. 1991. *Prime time activism*. Boston: South End Press.

Sapien, J., and A. Kessler. 2010. Habitat for Humanity to look at drywall. *ProPublica*, July 1. http://www.propublica.org/article/habitat-for-humanity-headquarters-to-look-at-defective-drywall (accessed July 1, 2010).

Scahill, J. 2010. BP and US government "Command Center" guarded by company from Afghan Embassy hazing scandal. *Nation*, May 28. http://www.thenation.com/blog/bp-and-us-government-command-center-guarded-company-afghan-embassy-hazing-scandal (accessed May 28, 2010).

Schmitt, E., and D. Johnston. 2008. States chafing at U.S. focus on terrorism. *New York Times*, May 26. http://www.nytimes.com/2008/05/26/us/26terror.html?hp (accessed May 26, 2008).

Schnailberg, A. 1980. *The environment: From surplus to scarcity*. New York: Oxford University Press.

Schneider, A., and D. McCumber. 2004. *An air that kills: How the asbestos poisoning of Libby, Montana, uncovered a national scandal*. New York: Berkley Books.

Schrope, M. 2010. Upbeat oil report questioned: Researchers see major uncertainties in Deepwater Horizon spill assessment. *Nature*, August 10. http://www.nature.com/news/2010/100810/full/466802a.html?s=news_rss (accessed August 10, 2010).

Schudson, M. 1995. *The power of news*. Cambridge, MA: Harvard University Press.

Schwartz, N. 2010. BP played key role in botched *Exxon Valdez* response. *Huffington Post*, May 25. http://www.huffingtonpost.com/2010/05/25/bp-exxon-valdez-response-gulf-oil-spill_n_588335.html (accessed May 25, 2010).

Schwartz, N., and M. Brown. 2010. Gulf oil spill sickness: Cleanup workers experience health problems, complain of flulike symptoms. *Huffington Post*, June 3. http://www.huffingtonpost.com/2010/06/03/gulf-oil-spill-sickness-c_n_598816.html (accessed June 3, 2010).

References

Scott, A. 1993. Interview with author. Lerwick, Shetland Islands, February 14.

Seattle Post-Intelligencer. 2008. FEMA: Stubborn facts. *Seattle Post-Intelligencer Editorial Board*, February 17. http://seattlepi.nwsource.com/opinion/351522_femaed.html (accessed February 19, 2008).

Serrano, R. 2010. Feds weigh a criminal probe of BP. *Los Angeles Times*, May 28. http://www.latimes.com/news/nationworld/nation/la-na-oil-spill-investigation-20100529,0,3427456.story (accessed May 28, 2010).

Sheppard, K. 2010a. Obama's sluggish oil spill response: Why has the administration been so slow to take charge of the disaster in the Gulf? *Mother Jones*, May 25. http://motherjones.com/politics/2010/05/obamas-sluggish-response-oil-spill (accessed May 25, 2010).

———. 2010b. Palin blames BP spill on "extreme enviros." *Mother Jones*. June 2. http://motherjones.com/blue-marble/2010/06/palin-blames-bp-spill-enviros (accessed June 2, 2010).

———. 2010c. NOAA report on amount of oil in Gulf won't be released for months. *Huffington Post*, August 19. http://www.huffingtonpost.com/2010/08/19/gulf-oil-spill-full-noaa-_n_687531.html (accessed August 19, 2010).

Sheppard, P. 1990. Interview with author. August 17, 1990.

Sherman, S. P. 1989. Smart ways to handle the press: Reporters are strange beasts, and as Exxon's ordeal shows, dealing with them can be tough. You've got to face facts—and whatever you do, you've got to talk. *CNN.com*, June 19.

Sherrill, R. 1983. *The oil follies of 1970-1980: How the petroleum industry stole the show (and much more besides)*. Garden City, NY: Anchor Press/Doubleday.

Shetland Islands Council. 1993a. Press release. Lerwick, January 27.

———. 1993b. Public health advice (concerning "water supply"; "crops"; "exposure to airborne pollution"; and "personal respiratory protection"). Lerwick, n.d.

———. 1993c. Public health advice. General statement issued to the public. Lerwick, January 29.

———. 1993d. Public notice. Lerwick, January 10.

———. 1993e. Public notice: Action to take to minimize skin contact with oil deposits. Lerwick, January 16.

———. 1993f. The use of dispersants. Lerwick, n.d.

Shrader-Frechette, K. S. 1991. *Risk and rationality*. Berkley: University of California Press.

Shrivastava, P. 1987. *Bhopal: Anatomy of a crisis*. Cambridge, MA: Ballinger Publishing Company.

Sierra Club. n.d. Press release: Centers for Disease Control study confirms dangers of FEMA trailers announcement comes nearly two years after Sierra Club first raised issue. http://action.sierraclub.org/site/ (accessed July 21, 2009).

———. 2008. Toxic trailers: Tests reveal high formaldehyde levels in FEMA trailers. April. www.toxictrailers.com. (accessed April 20, 2008)

References

———. 2009. Toxic trailers. Sierra Club, Mississippi Chapter. January 4. http://mississippi.sierraclub.org/issues/ (accessed July 21, 2009).

Simpson, H. 2010. Interview with author. Conducted by phone, June 13.

Sitter, C. 1989. *All things considered*. Interviewed by Noah Adams. May 19.

Skipp, C. 2008. Toxic trailers: Hurricane Katrina's victims cope with yet another ordeal— unhealthy residences provided by Uncle Sam. *Newsweek*, February 16. http://www.newsweek.com/id/112828 (accessed February 18, 2008).

Slovic, P. 1987. Perception of risk. *Science* 236: 280–85.

Smith, L. 1990. Interview with author. Homer, Alaska, August 1.

Smith, S. 2009. Months after ash spill, Tennessee town still choking. *CNN.com*, July 13. http://www.cnn.com/2009/HEALTH/07/13/coal.ash.illnesses/ (accessed July 14, 2009).

Sohn, P. 2009. Future murky for ash disposal. *Chattanooga Times Free Press*, June 29. http://timesfreepress.com/news/2009/jun/29/future-murky-for-ash-disposal/ (accessed June 29, 2009).

Solomon, G. M., M. Hjelmroos-Koski, M. Rotkin-Ellman, and S. K. Hammond. 2006. Airborne mold and endotoxin concentrations in New Orleans, Louisiana after flooding, October–November 2005. *Environmental Health Perspectives Online,* June 12. http://dx.doi.org (accessed June 12, 2006).

Spake, A. 2007. Dying for a home: Toxic trailers are making Katrina refugees ill. *Nation*, February 15. http:www.alternet.org/module/printversion/48004 (accessed September 13, 2007).

Spence, H. 1989a. Fertilizer blamed for illnesses. *Homer News*. August 24. Section 1A.

———. 1989b. Seldovians charge worker's health neglected. *Homer News*. August 31. Section 1A.

———. 1990a. Lujan: Mother Nature has done a lot. *Homer News*. April 12. Section 1A.

———. 1990b. Was spill 38 million gallons? Exxon figure disputed by biologist. April 12. Section 1A.

Spencer, P. 1990. *White silk, black tar: A journal of the Alaska oil spill*. Minneapolis: Bergamot Press.

Spill the Truth. n.d. Facts on Atlantis, BP and the administration. Spill the Truth: A Project of Food and Water Watch. http://www.spillthetruth.org/category/facts/ (accessed June 29, 2010).

St. Bernard Parish Community Member. 2006. Interview with author. St. Bernard Parish, Louisiana, March 15.

St. Bernard Parish Homeowner. 2006. Interview with author. St. Bernard Parish, Louisiana, March 18.

St. Bernard Parish Resident. 2006. Interview with author. St. Bernard Parish, Louisiana, March 18.

———. 2006. Interview with author. St. Bernard Parish, Louisiana, April 15.

References

Stallings, R. A. 1995. *Promoting risk: Constructing the earthquake threat.* New York: Aldine de Gruyter.

Stauber, J., and S. Rampton. 1995. *Toxic sludge is good for you: Lies, damn lies and the public relations industry.* Monroe, ME: Common Courage Press.

Stein, S. 2010a. 'Rand Paul: Obama sounds un-American' for criticizing Gulf oil spill. *Huffington Post,* May 21. http://www.huffingtonpost.com/2010/05/21/rand-paul-obama-sounds-un_n_584661.html (accessed May 21, 2010).

———. 2010b. Haley Barbour: Oil? What oil? Press should stop scaring tourists. *Huffington Post,* June 6. http://www.huffingtonpost.com/2010/06/06/haley-barbour-oil-what-oi_n_602088.html (accessed June 6, 2010).

Stephens, S. 2002. Bounding uncertainty: The post-Chernobyl culture of radiation protection experts. In *Catastrophe & Culture: The Anthropology of Disaster,* edited by S. M. Hoffman and A. Oliver-Smith. Santa Fe: School of American Research Press.

Stockpole, J. 1990. Interview with author. Homer, Alaska, June 7.

Stout, M. 1990. Interview with author. Homer, Alaska, July 14.

Stranahan, S. 2003. Air of uncertainty. *American Journalism Review.* January–February.

Sullivan, E. 2008. Draft plan: FEMA may use trailers in new disaster. *Associated Press,* June 3. http://ap.google.com/article/ (accessed June 4, 2008).

Swan Pond Road Resident. 2009. Interview with author. Roane County, Tennessee, February.

Tatitlek Woman. 1990. Interview with author. Tatitlek, Alaska, June 13.

Taylor, M. 2010. Since spill, feds have given 27 waivers to oil companies in gulf. *McClatchy Washington Bureau,* May 7. http://www.mcclatchydc.com/2010/05/07/93761/despite-spill-feds-still-giving.html (accessed May 8, 2010).

Taylor, M., and E. Bolstad. 2010. BP "systemic failure" endangers Gulf cleanup workers. *McClatchy Washington Bureau,* May 28. http://www.mcclatchydc.com/2010/05/28/95049/bps-systemic-failure-harms-workers.html (accessed May 28, 2010).

Taylor, M., and R. Schoof. 2010. BP withholds oil spill facts—and government lets it. *McClatchy Washington Bureau,* May 17. http://www.mcclatchydc.com/2010/05/18/94415/bps-secrecy-keep-facts-on-gulf.html (accessed on May 18, 2010).

Technology Review. 1976. *Offshore oil: The threat and the promise.* Cambridge: Massachusetts Institute of Technology.

Tellus Institute. 2003. "Daubert: The most influential Supreme Court decision you've never heard of." A Publication of the Project on Scientific Knowledge and Public Policy, coordinated by the Tellus Institute. June. http://www.defendingscience.org/upload/Daubert-The-Most-Influential-Supreme-Court-Decision-You-ve-Never-Heard-Of-2003.pdf.

References

Tennessee Department of Health. 2009a. Coal fly ash release. Fact sheet. Tennessee Department of Health. February 13.

————. 2009b. Public Health Assessment. "Tennessee Valley Authority (TVA) Kingston Fossil Plant Coal Ash Release." Roane County, Tennessee. December 9.

Tennessee Valley Authority. 2009. Inspection report: Review of the Kingston Fossil Plant ash spill root cause study and observations about ash management. *Tennessee Valley Authority, Office of the Inspector General*. 2008-12283-02. July 23.

Thomas. A. 2010. EPA whistleblower—Public can't handle the truth regarding Corexit 9500. Intel Hub, July 28. http://theintelhub.com/2010/07/28/epa-whistle blower-public-cant-handle-the-truth-regarding-corexit9500/ (accessed July 28, 2010).

Tilove, J. 2010a. BP, federal officials differ over best dispersant to use in Gulf. *Times-Picayune*, May 22. http://www.nola.com/news/gulf-oil-spill/index.ssf/2010/05/bp_federal_officials_differ_ov.html (accessed May 22, 2010).

————. 2010b. Drilling critics hammer at oil execs over spill emergency plans. *Times-Picayune*, June 16. http://www.nola.com/news/t-p/frontpage/index.ssf?/base/news-14/1276669937112120.xml&coll=1 (accessed June 16, 2010).

Tilove, J., and B. Alpert. 2010. Salazar not budging on ending drilling ban early. *Times-Picayune*, June 10. http://www.nola.com/news/t-p/frontpage/index.ssf?/base/news-14/1276151408112170.xml&coll=1 (accessed June 10, 2010).

Times-Picayune. 2008. Editorial: Unsafe havens. *Times-Picayune*, February 17. http://www.nola.com (accessed February 17, 2008).

————. 2010a. BP shares fault in deepwater horizon explosion: An editorial. *Times-Picayune*, May 27. http://www.nola.com/news/gulf-oil-spill/index.ssf/2010/05/gulf_of_mexico_rig_explosion_i.html (accessed May 27, 2010).

————. 2010b. Four oil-cleanup workers fall ill; Breton Sound fleet ordered back to dock. *Times-Picayune*, May 26. http://www.nola.com/news/gulf-oil-spill/index.ssf/2010/05/four_oil-cleanup_workers_fall.html (accessed May 26, 2010).

Titchenor, J., and W. Ross. 1974. Acute or chronic stress as determinants of behavior, character, and neurosis. In *American Handbook of Psychiatry*, edited by A. Arieti and E. Brody, 39–60. New York: Basic Books.

Totemoff, D. 1990. Interview with author. Homer, Alaska, April 11.

Totemoff, J. 1990. Interview with author. Homer, Alaska, April 11.

Totemoff, P. 1990. Interview with author. Homer, Alaska, April 11.

Toxic Culture. n.d. FEMA trailers: Death capsules for the poor. *ToxicTrailers.com*. http://www.toxictrailers.com (accessed July 21, 2009).

Tuchman, G. 1978. *Making news: A Study in the construction of reality*. New York: Free Press.

Turner et al. v Murphy Oil USA, Inc. 2006. 05-4206 (Eastern District Court of Louisiana, 2006).

References

Tuscaloosa News. 2009. Coal ash dump site in Alabama not welcome. July 6. http://www.tuscaloosanews.com/article/20090706/NEWS/907059973 (accessed July 6, 2009).

U.S. Bureau of the Census. n.d. State and county QuickFacts: Perry County, Alabama. http://quickfacts.census.gov/qfd/states/01/01105.html (accessed June 26, 2009).

U. S. Coast Guard. 1993. T/V EVOS: Federal on-scene coordinators report. Washington, DC: U. S. Department of Transportation.

U. S. Congress. House. Committee on Energy and Commerce. 2010. *Memorandum to members of the subcommittee on oversight and investigations and subcommittee on energy and environment from majority staff.* 111th Congress, Second Session, July 19.

U.S. Congress. House. Committee on Energy and Commerce, Subcommittee on Commerce, Trade, and Consumer Protection. 2010. Hearing on: *Public sales of Hurricane Katrina/Rita FEMA trailers: Are they safe or environmental time bombs.* 111th Congress, Second Session, April 28.

U.S. Congress. House. Committee on Natural Resources, Subcommittee on Energy and Mineral Resources. 2008. *Testimony of the Honorable Earl E. Devaney, inspector general for the Department of the Interior.* 110th Congress, Second Session, March 11.

———. 2010. *Testimony of Mary L. Kendall, acting inspector general for the Department of the Interior.* 111th Congress, Second Session, June 17.

U.S. Congress. House. Committee on Oversight and Government Reform. 2007a. *Prepared remarks of James D. Harris, Jr., hearing before the Committee on Oversight and Government Reform.* 110th Congress, Second session, July 19.

———. 2007b. *Prepared remarks of Paul Stewart, hearing before the Committee on Oversight and Government Reform.* 110th Congress, Second session, July 19.

———. 2007c. *Prepared testimony of Lindsay Huckabee, hearing before the Committee on Oversight and Government Reform.* 110th Congress, Second session, July 19.

———. 2007d. *Hearing summary: Hearing on FEMA's toxic trailers, testimony of Scott Needle, MD, FAAP, on behalf of the American Academy of Pediatrics.* 110th Congress, Second session, July 19.

———. 2007e. *The serious public health issues resulting from formaldehyde exposures within FEMA travel trailers issued to hurricane disaster victims, and recommended action items. Testimony of Mary C. DeVany before the Committee on Oversight and Government Reform.* 110th Congress, Second session, July 19.

———. 2008a. *Chairman Waxman expands toxic trailers investigation.* Hurricane Katrina Response Statement. 110th Congress, Second session, February 14. http://www.oversight.house.gov/story.asp?ID=1751 (accessed June 21, 2009).

———. 2008b. *Committee holds hearing on manufacturers of FEMA toxic trailers.* Hurricane Katrina Response Statement. 110th Congress, Second session, July 9. http://www.oversight.house.gov/story.asp?ID=2069 (accessed June 21, 2009).

References

———. 2008c. *Committee holds hearing on manufacturers of FEMA toxic trailers: Chairman Waxman's opening statement.* Hurricane Katrina Response Statement. 110th Congress, Second session, July 9. http://www.oversight.house.gov/story.asp?ID=2073 (accessed June 21, 2009).

———. 2008d. *CDC Congressional testimony: Statement of Michael McGeehin, PhD, MSPH.* 110th Congress, Second session, July 9.

———. 2008e. *Trailer manufacturers and elevated formaldehyde levels.* 110th Congress, Second session, July 9.

U.S. Congress. House. Committee on Science and Technology. 2008a. *Press release: Subcommittees demand explanation for attempts to control report on formaldehyde in trailers housing hurricane victims.* 110th Congress, Second session, January 28.

———. 2008b. *Miller/Lampson letter to the director of the National Center for Environmental Health regarding ATSDR testing of Hurricane Katrina/Rita FEMA trailers.* 110th Congress, Second Session, January 28.

U.S. Congress. House. Committee on Science and Technology, Subcommittee on Investigations and Oversight. 2008a. *Response to concerns of formaldehyde in FEMA travel trailers: Testimony of Heidi Sinclair MD, MPH, FAAP.* 110th Congress, Second Session, April 1.

———. 2008b. *Toxic trailers—toxic lethargy: How the Centers for Disease Control and Prevention has failed to protect the public heath, majority staff report.* 110th Congress, Second Session, September.

———. 2008c. *Toxic trailers: Have the Centers for Disease Control failed to protect the public? Response to Issues 1–4: Dr. Christopher T. De Rosa, NCEH/ATSDR.* 110th Congress, Second Session, April 1.

———. 2008d. *Toxic trailers: Have the Centers for Disease Control failed to protect the public? Opening statement by Chairman Bart Gordon.* 110th Congress, Second Session, April 1.

———. 2008e. *Toxic trailers: Have the Centers for Disease Control failed to protect the public? Opening statement by Chairman Brad Miller.* 110th Congress, Second Session, April 1.

———. 2008f. *Toxic trailers: Have the Centers for Disease Control failed to protect the public? Testimony of Becky Gillette.* 110th Congress, Second Session, April 1.

———. 2008g. *Toxic trailers: Have the Centers for Disease Control failed to protect the public? Testimony of Dr. Meryl H. Karol.* 110th Congress, Second Session, April 1.

———. 2009. "The Agency for Toxic Substances and Disease Registry (ATSDR): Problems in the past, potential for the future." Report by the majority staff of the Subcommittee on Investigations and Oversight of the Committee on Science and Technology, U.S. House to Subcommittee Chairman Brad Miller. 111th Congress, First Session, March 10.

References

U.S. Congress. House. Committee on Transportation and Infrastructure, Subcommittee on Water Resources and the Environment. 2009. *Statement of Richard W. Moore inspector general, Tennessee Valley Authority.* 111th Congress, First Session, July 28.

U.S. Congress. Senate. Committee on Commerce, Science, and Transportation. 1989. *Exxon -Valdez oil spill and its environmental and maritime implications.* Washington, DC, 101st Congress, First Session, April 6.

U.S. Congress. Senate. Committee on Environment and Public Works, Subcommittee on Wildlife and Water. 2010. *Testimony of Stanley Senner, Ocean Conservancy "Assessing natural resource damages following the BP Deepwater Horizon disaster."* July 27.

U.S. Congress. Office of Technology Assessment. 1990. *Coping with an oiled sea.* Background paper. Washington, DC: U.S. Government Printing Office.

U.S. Department of Homeland Security, Office of the Inspector General. 2009. FEMA response to formaldehyde in trailers, redacted. OIG-09-83. June.

U.S. Environmental Protection Agency. 2001a. Whitman details ongoing agency efforts to monitor disaster sites, contribute to cleanup efforts. Press release, September 18. http://www.epa.gov/wtc/stories/headline_091801.htm (accessed September 18, 2010).

———. 2003b. Evaluation report: EPA's response to the World Trade Center collapse: Challenges, successes, and areas for improvement. *Office of Inspector General.* Report no. 2003-P-00012: August, 21.

———. 2003c. Evaluation report survey of air quality information related to the World Trade Center collapse. *Office of Inspector General.* Report no. 2003-P-00014: September 26.

———. n.d. Environmental justice: Basic information. http://www.epa.gov/compliance/basics/ejbackground.html (accessed July 15, 2009).

———. 2010a. EPA: BP must use less toxic dispersant. Press release, May 20. http://yosemite.epa.gov/opa/admpress.nsf/d0cf6618525a9efb85257359003fb69d/0897f55bc6d9a3ba852577290067f67f!OpenDocument (accessed on May 20, 2010).

———. 2010b. EPA's toxicity testing of dispersants. http://www.epa.gov/bpspill/dispersants-testing.html (accessed May 23, 2010).

U.S. Government Accountability Office (GAO). 2007. Hurricane Katrina: EPA's current and future environmental protection efforts could be enhanced by addressing issues and challenges faced on the Gulf Coast. Report to Congressional Committees. Washington, DC. June.

U.S. Government Federal Background Sheet. 1991. Federal agencies release preliminary assessment of injuries relating to the *Exxon Valdez* oil spill.

Unnamed Journalist. 2009. Interview with author. Roane County, Tennessee, January 12.

References

Urbina, I. 2010a. Banned trailers return for latest Gulf disaster. *New York Times*, June 30. http://www.nytimes.com/2010/07/01/us/01trailers.html?_r=1&scp= 1&sq=Banned%20trailers%20return%20for%20latest%20Gulf%20&st=cse (accessed June 30, 2010).

———. 2010b. BP is pursuing Alaska drilling some call risky. *New York Times*, June 23. http://www.nytimes.com/2010/06/24/us/24rig.html?scp=1&sq=BP%20is %20pursuing%20Alaska%20drilling%20some%20call%20risky&st=cse (accessed June 23, 2010).

———. 2010c. Conflict of interest worries raised in spill tests. *New York Times*, May 20. http://www.nytimes.com/2010/05/21/science/earth/21conflict.html?fta=y (accessed May 20, 2010).

———. 2010d. Documents show early worries about safety of rig. *New York Times*, May 29. http://www.nytimes.com/2010/05/30/us/30rig.html?scp=1&sq=%22Documents %20show%20early%20worries%20about%20safety%20of%20rig%22&st=cse (accessed May 29, 2010).

———. 2010e. In Gulf, it was unclear who was in charge of oil rig. *New York Times*, June 5. http://www.nytimes.com/2010/06/06/us/06rig.html?scp=1&sq=%22in%20Gulf,% 20it%20was%20unclear%22&st=cse (accessed June 5, 2010).

———. 2010f. Inspector general's inquiry faults regulators. *New York Times*, May 24. http://www.nytimes.com/2010/05/25/us/25mms.html?scp=1&sq=Inspector% 20General's%20inquiry%20faults%20regulators&st=cse (accessed May 24, 2010).

———. 2010g. U.S. said to allow drilling without needed permits. *New York Times*, May 13. http://www.nytimes.com/2010/05/14/us/14agency.html (accessed May 13, 2010).

van Eijndhoven, J. 1994. Disaster prevention in Europe. In *Learning from disaster*, edited by S. Jasanoff, 113–32. Philadelphia: University of Pennsylvania Press.

Vidal, J. 2010. Nigeria's agony dwarfs the Gulf oil spill. The U.S. and Europe ignore it. *CommonDreams.org*, May 30. http://www.commondreams.org/headline/2010/ 05/30-0 (accessed May 30, 2010).

Volunteer. 1990. Interview with author. Homer, Alaska, July 3.

Wald, M. 2010. Despite directive, BP used dispersant often, panel finds. *New York Times*, August 1.

Walker, T. 2008. FEMA faces lawsuits over "toxic" trailers. *WTHR.com*, May 27. http://www.wthr.com/global/story.asp?s=8385568 (accessed May 28, 2008).

Walker-Journey, J. 2009. EPA approves TVA's bid to store recovered coal ash in Alabama. July 6. Beasley, Allen, Crow, Methvin, Portis & Mile, P.C. http://www. coal-ash-spill.com/news/2009/07/06epa-approves-tvas-bid-to-store-recovered-coal-ash-in-alabama/ (accessed July 15, 2009).

Wallace, A. 1987. *St. Clair: A nineteenth century coal town's experience with a disaster-prone Industry*. Ithaca, NY: Cornell University Press.

Wallace, A. P. 1956. *Human behavior in extreme situations: A survey of literature and suggestions for further research.* Washington, DC: National Research Council.

Wallack, L., L. Dorfman, D. Jernigan, and M. Themba, eds. 1993. *Media Advocacy and Public Health.* London: Sage Publications.

Wang, M. 2010a. Coast Guard flagged potential problems with spill response in 2004. *ProPublica*, May 21. http://www.propublica.org/ion/blog/item/coast-guard-flagged-potential-problems-with-spill-response-since-2004 (accessed May 21, 2010).

———. 2010b. Health effects after *Exxon Valdez* went unstudied. *WKRG News*, June 30. http://www.wkrg.com/gulf_oil_spill/article/health-effects-after-exxon-val-dez-went-unstudied/902123/Jun-30-2010_12-43-pm/ (accessed June 30, 2010).

Wardell, J. 2010. Gulf oil spill causes other nations to rethink drilling. *Huffington Post*, June 20. http://www.huffingtonpost.com/2010/06/20/gulf-oil-spill-causes-oth_n_618881.html (accessed June 21, 2010).

Washington Post. 2008. Those FEMA trailers: Help finally comes to those who have suffered enough. February 22. http://www.washingtonpost.com/wp-dyn/content/article/2008/02/21/AR2008022102553_pf.html (accessed February 23, 2008).

Watson, S. 2004. Interview with author. Austin, Texas, March 15.

Waxman, H., and B. Rush. 2010. Letter to M. N. Johnson and W. C. Fugate, July 6. http://energycommerce.house.gov/documents/20100706/JohnsonFugate.2010.7.6.pdf (accessed July 6).

Waxman, H., B. Stupak, and F. Pallone. 2010. Letter to R. Tillerson, July 1. http://energycommerce.house.gov/documents/20100701/Tillerson.ExxonMobil.2010.7.1.pdf (accessed July 1).

Wearden, G. 2010. Gulf oil spill: Lacked the right tools to deal with crisis, chief executive admits. *Guardian.co.uk*, June 3. http://www.guardian.co.uk/business/2010/jun/03/gulf-oil-spill-senators-ask-bp-to-suspend-dividends (accessed June 3, 2010).

Webb, T. 2010. BP boss admits job on the line over Gulf oil spill. *Guardian.co.uk*, May 14. http://www.guardian.co.uk/business/2010/may/13/bp-boss-admits-mistakes-gulf-oil-spill (accessed May 14, 2010).

Webb, T., and E. Pilkington. 2010. Gulf oil spill: BP could face ban as U.S. launches criminal investigation. *Guardian.co.uk*, June 2. http://www.guardian.co.uk/environment/2010/jun/01/gulf-oil-spill-bp-future (accessed June 2, 2010).

Weber, H. R. 2010. Stupak: BP won't let Congress talk to key employees. *Huffington Post*, June 26. http://www.huffingtonpost.com/2010/06/25/stupak-bp-employ-ees-congress_n_626416.html (accessed June 26, 2010).

Wedel, J. R. 2009. *Shadow elite: How the world's new power brokers undermine democracy, government, and the free market.* New York: Basic Books.

Weick, K. 1976. Educational organizations as loosely coupled systems. *Administrative Science Quarterly* 21, no. 1: 1–19.

References

White, M. 1993. Interview with author. Lerwick, Shetland Islands, February 14.

White, M. 2008a. CDC urges action on hazardous FEMA trailers. *BuzzFlash.org*, July 3. http://blog.buzzflash.com (accessed July 21, 2008)

——. 2008b. Trapped in toxic trailers by FEMA and profiteering industry. Katrina survivors victimized once again by Bush. *BuzzFlash.org*, July 9. http://blog. buzzflash.com/ (accessed June 21, 2009).

Whiteside, T. 1979. *The pendulum and the toxic cloud: The course of dioxin contamination*. Boston: Yale University Press.

Whitney, D. 1990. Birds killed to make case against Exxon. *Anchorage Daily News*, October 18, A1 and A10.

Whittaker, J. 1993. Interview with author. Lerwick, Shetland Islands, February 15.

Wills, J. 1991. *A place in the sun: Shetland and oil-myths and realities*. Edinburgh, Scotland: Mainstream Publishing.

Wing, N. 2010. John Culberson: Gulf oil spill is a "statistical anomaly," Texas congressman says. *Huffington Post*, June 10. http://www.huffingtonpost. com/2010/06/09/john-culberson-gulf-oil-s_n_606443.html (accessed June 10, 2010).

Winslow, R. 1978. *Hard aground: The story of the Argo Merchant oil spill*. New York: W. W. Norton and Company.

Wolf, E. 1990. Facing power: Old insights, new questions. *American Anthropologist* 92, no. 3: 586–96.

WTVC News Channel 9. 2009. Massive coal ash spill still affecting residents. June 24. http://www.newschannel9.com/news/says-979486-tva-turpin.html (accessed June 25, 2009).

Yergin, D. 1992. *The prize. The epic quest for oil, money, and power*. New York: Simon and Schuster.

Young, A. 2008a. CDC discussed Katrina trailer risk a year ago: Panel examines whether FEMA, CDC worked together to downplay risks. *Atlanta Journal-Constitution*, February 15. http://www.ajc.com/news/content/news/stories/2008/02/14/cdc-side_0215.html (accessed February 17, 2008).

——. 2008b. Katrina report slams CDC. Investigators: Agency failed to protect victims from fumes. *Atlanta Journal-Constitution*, April 1. http://www.ajc. com/news/content/news/stories/2008/03/31/cdckatrina_0401.html (accessed April 4, 2008).

Young, A. L. 2005. Alive in Truth. http://www.aliveintruth.org/a_pages/who_we_ are.htm (accessed September 11, 2010).

Zarembo, A. 2003. Column one: Funding studies to suit need. *Los Angeles Times*. December 3:1A, 15A.

Index

Index

Index

Index

diesel fumes, 37
formaldehyde, 123
polynuclear aromatic hydrocarbons, 35
radiation, 12–13
Shetland Islands oil spill and, 80
St. Bernard Parish air quality and, 96
cardiac disease, 95
Cato Institute, 164
cattle
    PBB contamination in Michigan, 48, 49, 56
    Shetland Islands oil spill and, 74, 82–83
CDC. *See* Centers for Disease Control and Prevention
censorship of science
    by BP, 229
    by Exxon, 41
    by G. W. Bush administration, 165
Center for Toxicology and Environmental Health (CTEH)
    Deepwater oil spill and, 233
    Murphy Oil spill and, 99, 100
Centers for Disease Control and Prevention (CDC). *See also* Agency for Toxic Substances and Disease Registry
    FEMA trailers and, 111
    formaldehyde toxicity levels set by, 120
    Vernon Houk and, 203
central nervous system effects, 36, 230
cerebral palsy, 133
Chalmette, Louisiana
    cancer rates in, 93–94
    civic pride in, 92
    industrialization of, 94–95
    inundation of, 91
Chemical Manufacturers Association, 203, 204
Cheney, Dick, 196, 237
Cherniack, Martin, 157
Chernobyl nuclear accident
    animal casualties, 48, 56
    science seen as cause of, 168

chest pains
    Deepwater oil spill and, 230
    Shetland Islands oil spill and, 77
Chevron
    activities in Amazon, 185
    drilling platform fire, 212
    Hurricanes Katrina/Rita and, 211
children
    animals as, 57
    FEMA trailers and, 112–113, 119
    indigenous people as, 58
    Murphy Oil spill and, 98–99, 99, 100
    Shetland Islands oil spill and, 71, 75, 79
    TVA ash spill and, 130, 137
Chilean earthquake, 153
chitons, 22
Christian theology, 48
chromium, 132, 134, 145
citizen's science, 169
Civil Action, A (Harr), 150, 186
Clarke, Lee, 147, 152, 215, 216–217
Clean Air Act, 96
cleanup workers
    after 9/11, 198–200, 202
    Deepwater oil spill: FEMA trailers, 236; health concerns, 229–235; parallels with *Exxon-Valdez* workers, 246; safety training, 232
    Exxon-Valdez oil spill: animal casualties and, 52, 53; health and safety concerns, 34–38; safety training, 35, 36
    Murphy Oil spill, 97
    Shetland Islands oil spill and, 75
    TVA ash spill, 133, 136–137
climate change, 164–165, 177
clinical monitoring. *See* health monitoring
Clyde, George, 240
CNN, 164
"Coal Ash Is More Radioactive than Nuclear Waste," 132
Coast Guard

Index

Index

Index

Index

Index

Index

Index

Index

Index

About the Author

Dr. Gregory Button is a nationally recognized expert on disasters who has been studying extreme events for over thirty years. As a reporter and producer for public radio, he covered and reported on the Three Mile Island nuclear accident, the controversy surrounding Love Canal, and the eruption of Mount St. Helens. As an academic he has conducted extensive field research on numerous other disasters, including the Exxon-Valdez oil spill, Hurricane Katrina and the TVA ash spill in eastern Tennessee. He is currently a faculty member in the Department of Anthropology at the University of Tennessee, Knoxville, and a Co-Director of the Center for the Study of Social Justice. Dr. Button also served as a U.S. Congressional Fellow in the Senate.